"I have been so many people":
A Study of Lee Smith's Fiction

Tanya Long Bennett

UNG
UNIVERSITY of
NORTH GEORGIA
UNIVERSITY PRESS

Dahlonega, GA

Copyright 2014, Tanya Long Bennett

All rights reserved. No part of this book may be reproduced in whole or in part without written permission from the publisher, except by reviewers who may quote brief excerpts in connections with a review in newspaper, magazine, or electronic publications; nor may any part of this book be reproduced, stored in a retrieval system, or transmitted in any form or by any means electronic, mechanical, photocopying, recording, or other, without written permission from the publisher.

Published by:
The University of North Georgia Press
Dahlonega, Georgia

Publishing Support by:
Lightning Source, Inc.
La Vergne, TN

Cover Art: "Origin" by Alisha Moss, 2013, hand-colored print, 24" x 36"
Cover Design: Jon Mehlferber and April Loebick

ISBN: 978-1-940771-07-6

Printed in the United States of America, 2014

For more information, please visit: http://www.ung.edu/university-press
Or e-mail: ungpress@ung.edu

Table of Contents

Acknowledgments V

Lee Smith in Context: An Introduction 1

Chapter One
Early Signs: *The Last Day the Dogbushes Bloomed*, *Something in the Wind*, and *Fancy Strut* 8

Chapter Two
The Drowning of Crystal Spangler in *Black Mountain Breakdown* 20

Chapter Three
Narrative Mourning: Textual Suspension of Past/Present in *Oral History* 30

Chapter Four
The Culminating Self in *Family Linen* 40

Chapter Five
The Protean Ivy in *Fair and Tender Ladies* 51

Chapter Six
"It was like I was *right there*": Primary Experience and the Role of Memory in *The Devil's Dream* 63

Chapter Seven
And the Word Was God: Narrative Negotiation of the Spirit/Flesh Split in *Saving Grace* 74

Chapter Eight
Always the Storyteller's Story: *The Last Girls* 87

Chapter Nine
"We are all just passing through": Contingency in *On Agate Hill* 97

Afterword 108

Acknowledgments

I offer my sincere thanks to Chuck Bennett, who keeps me engaged and sane; to Zach, Luke, and Tyler Bennett, who inspire and challenge me; to Donna and Philip Long, for giving me life, confidence, and compassion; to Brian Long and Perri Long Rosheger for their enduring friendship and support; to Margaret Bauer, for modeling what a tough, smart woman should be; to Donna Gessell, for being a tireless and insightful reader, as well as a good friend; and to B.J. Robinson, for convincing me that creativity is as necessary for survival as oxygen.

Lee Smith in Context: An Introduction

Oh, Joli, you get so various as you get old! I have been so many people. And yet I think the most important thing is Don't forget. Don't ever forget. I tell you this now in particular. A person cannot afford to forget who they are or where they came from, or so I think, even when the remembering brings pain. (*Fair and Tender Ladies* 266)

Among Lee Smith's readers there seems to be a common fascination with her charismatic character Ivy Rowe, whose distinctive voice, conveyed in the above passage, so thoroughly captivates readers. With Ivy, protagonist of the novel *Fair and Tender Ladies*, Smith accomplishes so much: an authentic and provocative narrative voice, a vivid sense of the mountains and their bearing on the experience and sensibility of their inhabitants, a rich exploration of epistolary expression, and the poetry inherent in the Appalachian idiom. This kind of literary prowess has convinced many readers of Smith's merits as a fiction writer and has inspired critics to consider her work in the context of authors such as William Faulkner, Toni Morrison, Gloria Naylor, and the Brontes.[1] Yet, this passage also reflects Smith's deep and consistent exploration of the *self*, of personal identity in an ever-changing landscape and society.

Smith's body of work examines the influence of significant factors—such as place, memory, art, tradition, social expectation, media, religion, history, and story—on personal identity. Enriching her treatment of the subject, she explores this issue always with a consciousness of the self's

---

[1] See Kevin Massey, "'Wonderful Terms and Phrases': Contrasting Dialect in William Faulkner's *As I Lay Dying* and Lee Smith's *Oral History*"; Paula Gallant Eckard, *Maternal Body and Voice in Toni Morrison, Bobbie Ann Mason, and Lee Smith*; Joycelyn Hazelwood Donlon, "Hearing Is Believing: Southern Racial Communities and Strategies of Story-Listening in Gloria Naylor and Lee Smith"; H.H. Campbell, "Lee Smith and the Bronte Sisters."

ultimate indeterminability. For example, Smith says of the past as a possible clue to understanding ourselves,

> ...[Y]ou can never understand the past really. You can never understand why people behave as they do or what passions they pursue. The human heart is just an enormous mystery, I think; we can never understand it, and all you can do is form your new union, your own new ritual, your own new forms, and go ahead. What else are you going to do? Imperfect as it all is, you have to keep on going. (Qtd. in Rebecca Smith, "A Conversation with Lee Smith" 24)

In twelve novels and four collections of short stories, Smith draws us into a rigorous exploration of the self, the location and essence of which are often sought in a landscape of shifting and imagined markers.

Smith's consistent interest in identity targets a key element of human experience in the postmodern era, the psychological rootlessness that cannot be escaped once society has recognized that, as Jacques Derrida described in *Of Grammatology* (1967), reality is constructed to a great extent by language and, as a result, must be held suspect. As Smith's work reveals, the challenges of this postmodern consciousness can be portrayed with particular impact in the context of a historically idealized South. Perceiving the world through a lens of Southern (often Appalachian) cultural ideology, Smith's characters find themselves caught between the entrapping/seemingly stable notions of what they "ought" to be and the liberated/rootless existence outside these notions. This state is distinctly characteristic of the postmodern subject; modernist characters such as William Faulkner's Joe Christmas (*Light in August*), for example, never get far enough beyond restrictive Southern ideology to discover the disorientation that comes with being liberated from that ideology.

In *The Southern Writer in the Postmodern World*, Fred Hobson draws distinctions among the major twentieth century waves of Southern writers. He notes of those whose work began to appear in the 1970s and '80s, including Lee Smith, that their relationship to the past is significantly different from that of earlier writers. Hobson asserts that unlike generations before them, who wrote to come to terms with the guilt and defeat of their Southern past, the contemporary authors "immerse their characters in a world of popular or mass culture, and their characters' perceptions of place, family, community, and even myth are greatly conditioned by popular culture, television, movies, rock music, and so forth" (10). Notably, in responding to her literary past, Smith does not reiterate the position of her well-recognized predecessors, whose power, according to Hobson, "stemmed in part from a philosophical, even mildly didactic intent, the kind of writing associated with the novel of ideas, the novel of historical meditation, or the novel concerned with sweeping social change" (10). Rather, her position reflects the complex problems of viewing the world through a postmodern consciousness in a Southern landscape. Smith's approach to the issue of identity, the pursuit of self, is not allegorical: Her protagonists are not primarily symbols of a New South attempting to deal with its past. Rather, they are individuals negotiating their world by way of their own perceptions and never quite getting at it.

Mark Royden Winchell explains that the Southern Renascence of the 1920s and '30s occurred at a time when conditions were ripe for the shaping of a coherent cultural Southern identity: "[T]he insular subculture that had been the traditional South was in the process of being assimilated into the mainstream of American life. It is precisely at this moment of transition when we are able to see a culture most clearly" (x). Winchell notes that while this self-perception was sustained from the time of World War I to mid-century, by the 1950s,

> *the Southern Renascence, as defined by [Allen] Tate, had pretty much run its course...It was only a matter of time before revisionist critics, who were motivated in part by their opposition to the social vision of the Agrarians, began to question whether the Renascence (and perhaps even the South itself) had ever really existed.* (xi)

Despite this opposition, the idealized South, characterized by "a strong sense of place, based on memory, insularity and a tragic sense of defeat in the Civil War" (Ladd 1629), survived into the second half of the twentieth century and is still pervasive in many commercial representations of the region. The values arising from this identity include "honor, chivalry towards women, gentleness with subordinates...and belief in evil, in the Fall" (Ladd 1629). Yet, in her thorough reflection of Southern literature's evolution over the last century, Barbara Ladd states,

> *The Souths of contemporary writers look a lot like the rest of the country: gated communities, McMansions, run-down housing projects where trash collects in the street, unresponsive bureaucracies. Historical fiction remains useful...but history seems to have loosened its grip somewhat on contemporary writers to make room for fictional reappraisals of the public and private events of the remembered past.* (1632)

As Ladd points out, for reasons of authenticity and social responsibility, critics such as Patricia Yaeger, Thadious M. Davis, and Suzanne W. Jones have begun to provide revised filters through which to understand literature of the South, which is much more multifaceted than it was treated in earlier decades.

In addition to the positive effects of this new perspective, however, it must be acknowledged that with the move away from the South's oversimplified "traditional" identity and toward a contemporary era characterized as the "information age," the postmodern Southern subject (like the postmodern subject in general) is often left feeling somewhat disoriented. Emphasizing the role of capitalism in this shift, Fredric Jameson argues that "[i]f the ideas of a ruling class were once the dominant (or hegemonic) ideology of bourgeois society, the advanced capitalist countries today are now a field of stylistic and discursive heterogeneity without a norm" (17). Of course advertisers are in the business of constantly conveying to us "the norm," perhaps providing the postmodern subject with some sense of a cultural orientation point; however, the clash of this "norm" with remnants of regional, religious, and historical values can cause a kind of fragmentation. One constantly negotiates among all the resulting versions of his or her self, attempting to retain a sense of identity that enables him/her to function. Like Ivy, the contemporary individual must, on some level, recognize that "I have been so many people."

The full body of Smith's work, both short stories and novels, reveals her willingness to take artistic risks in exploring this contemporary condition. As early as spring 2000, Jacqueline Doyle noted Smith's important contribution to letters:

> *In more than ten novels and short story collections, Smith has explored aspects of Southern life, history, and culture—its oral traditions, sacred and secular music, class divisions and race relations, mountain customs, Appalachian folklore, and the spiritual life and beliefs of white Southern fundamentalist churches.* (273)

In fiction that covers such a wide spectrum of human experience, Smith has yielded works that range from "a funny story," as *Fancy Strut* has been deemed (Buchanan 331), to "mythic" and "lyrical," as *Fair and Tender Ladies* has been described (Hill 104 and 107 respectively). In creating such a variety of narratives, Smith has revealed what many great writers do: that inherent in the process of artistic pursuit is the inevitability of producing some works that are more powerful than others. In the entire range of her fiction, however, Smith explores the concept of the self, the forces that shape it, and the never-ending process of trying to define it. In all of her stories, she raises provocative questions about identity.

Born in the small coal-mining mountain town of Grundy, Virginia, in 1944, Smith developed, early on, a keen interest in both oral history and its reflection/creation of its tellers and listeners. Her father owned a dime-store in Grundy, and as a child, Smith used her access to the store to glean stories from the townspeople as they did their shopping and talked to her father. Though she was not from the hollers herself, the folklore and the idiom of these stories made a strong impression on her imagination.

In the early 1960s, she enrolled at Hollins College in Roanoke, Virginia, and there studied creative writing under Louis D. Rubin, Jr., along with Annie Dillard, Anne Goodwyn Jones, and Lucinda Hardwick MacKethan.[2] Written during her senior year at Hollins, Smith's first novel, *The Last Day the Dogbushes Bloomed*, was published in 1968. *Something in the Wind* followed in 1971 and *Fancy Strut* in 1973. At work on short stories as well during this time, Smith published several between 1968 and 1981; two of these works, "Mrs. Darcy Meets the Blue-Eyed Stranger at the Beach" and "Between the Lines," won O. Henry awards in 1978 and 1980, respectively (Hill xiv). Smith's fiction was already receiving some positive critical attention by this time, but her most important work was yet to come.

In 1980, just before her short story collection *Cakewalk* appeared, Smith had a fourth novel published: *Black Mountain Breakdown*. Not only was this novel darker than her previous work, but it was also considered more seriously by critics. Anne Goodwyn Jones notes Smith's artistic growth evidenced in this work: "Her technical achievement reflects a growing conviction in Smith's fiction that character and time and place are inextricable" (261). Confirming Smith's interest in the forces that shape people, Katherine Kearns suggests that in this novel, Smith portrays the consequences of women's artistic self-denial: "Anxious to be 'normal'—to be beauty queens and majorettes and good wives—[Smith's early female characters] move toward the immobilizing truth of *Black Mountain Breakdown*: that to be all things to all people is to be nothing, to risk becoming a mirror woman whose last refuge is the shroud of catatonia" (175). In *Black Mountain Breakdown*, Smith had taken her fiction beyond adept storytelling to a level of philosophical and artistic investigation into the conditions of *self*.

Three years later, Smith produced *Oral History*, a work that is regarded by many as one of her most important novels. Anne Goodwyn Jones stated, not long after the book's appearance, "[T]he rich palpable texture, the complex humanity of point of view...are everywhere apparent in the

---

2 See Nancy C. Parrish's *Lee Smith, Annie Dillard, and the Hollins Group: A Genesis of Writers*.

tapestry of voices, oral and written, that composes Lee Smith's newest and finest fiction, *Oral History*" (265). Here, Jones reflects Smith's maturing use of perspective to better understand the fluidity of identity in its varying contexts. Although in 1985 *Oral History* was followed by *Family Linen*, a novel that did not quite reach its predecessor's level of sophistication, in 1988 Smith produced the novel often thought to be her finest, *Fair and Tender Ladies*. With this novel's memorable protagonist/narrator, Smith achieved a voice that has been praised as "healing," "lyrical," and "rhapsodic" (Hill 107). "Smith breathes into this novel," writes Dorothy Combs Hill, "the lyricism of a woman's life, gives it the dignity and durability of collective language—renders it mythic" (104). *Black Mountain Breakdown*, *Oral History*, and *Fair and Tender Ladies* had boosted Smith into the category of writers whose work is regarded for its influence on our self-perspectives, our society, and our culture.

In the past two decades, Smith has continued to produce new fiction at a steady pace. In 1990, she published her second collection of short stories, *Me and My Baby View the Eclipse* and, in the years after, produced *The Devil's Dream* (1992), *Saving Grace* (1995), the novella *The Christmas Letters* (1996), her third volume of short stories, *News of the Spirit* (1997), *The Last Girls* (2002), *On Agate Hill* (2006), and her most recent collection of short stories, *Mrs. Darcy and the Blue-Eyed Stranger* (2011). These works published since 1990 confirm Smith's ability to produce powerful narrative explorations of identity and its shaping factors.

This study examines Smith's novels, specifically their focus on characters seeking identity in a changing Southern landscape, both physical and imagined. Smith's short stories are fascinating as well, essentially delving into this same thorny issue. However, the scope of this book necessarily excludes the short fiction since the narrative strategies Smith uses in that genre deserve in-depth analysis in their own right. I hope to explore, in a separate study, Smith's contributions to short fiction. This study also excludes Smith's novelty novella, *The Christmas Letters*, since its shaping parameters are so different from those of her other novels. My study, then, focuses in particular on the other eleven novels, on the narrative strategies Smith employs there to reflect a significant contemporary human dynamic: that of continually imagining the self in conjunction with personal memory and with cultural constructions such as history, place, religion, media, and art.

In the chapters that follow, Smith's novels are examined to reveal her fascination with the *self* and her narrative strategies for exploring it. It is evidence of both Smith's artistic prowess and our obsession with the identity question that a broad array of contemporary theories can be employed to unveil Smith's insights on the issue. Of course, deconstructionists such as Jacques Derrida and Stanley Trachtenberg are essential in articulating the mid-twentieth century move away from Platonic notions of originality and centrality and toward the recognition that language itself is a tool in the constant and necessary task of generating meaning (as opposed to uncovering it). I rely, in this project, on the now widely-known precepts of deconstruction, and I engage Trachtenberg and Derrida specifically in Chapter 9 to frame the examination of *On Agate Hill* and Smith's conception of "home" in that novel.

But in other chapters, I have taken great liberty in applying the work of other theorists in cases where the approaches are fruitful in unraveling Smith's varied narrative strategies and effects. For example, in Chapter 3, Rodger Cunningham's discussion of "double alterity" is quite useful as a lens through which to understand *Oral History*. Of particular benefit is his explanation of the human tendency to seek a void or blank on which we may write the "other," simultaneously exoticizing and alienating the other. Chapter 3 applies this notion of "double alterity" to *Oral History* specifically since the narrative itself explores this blank and the dynamics set in motion by its being written upon. Similarly, Chapter 6 employs concepts basic to Zen, as represented by Thomas Merton and Alan W. Watts, to reveal the distinction in *The Devil's Dream* between primary and secondary experience. In this novel, Smith investigates the often problematic self-conscious postmodern experience of the world, the constant and somewhat tragic self-construction of the postmodern subject, and the resulting difficulty of achieving unselfconscious experience.

As a whole, this study examines what Smith's novels have brought her readership, noting, in the book's arrangement, the increased impact of her work as she has matured as a writer. As stated previously, Smith's fiction calls for more critical exploration. However, I hope what I offer here is an illuminating look at one of her central concerns. Smith's characters do not serve as philosophical mouthpieces, as have characters of many novels of the Western canon. Yet, through richly developed characters, carefully chosen narrative frameworks, compelling dialogue, and often haunting settings, she has rigorously and insightfully investigated the condition of the postmodern self, as I hope is evidenced in the following chapters.

Chapter One
Early Signs: *The Last Day the Dogbushes Bloomed*, *Something in the Wind*, and *Fancy Strut*

Lee Smith's *The Last Day the Dogbushes Bloomed* appeared in 1968, as the Southern Renascence was winding down and the Civil Rights Movement was leaving its indelible mark on the U.S. The timing of Smith's work alongside major changes occurring in U.S. culture is clearly evident in her fiction, even in the early novels. Although Smith did not respond to the political climate with "novels of ideas" or subversive historical novels, her characters are subject to the provocative questions and uncertainties characteristic of the post-Agrarian South and of postmodern America. In the second half of the twentieth century and the early twenty-first century, scholarship of Southern literature has consistently investigated this major social shift and how it affects our understanding of the literature as well as of Southern culture and identity. In particular, the challenge to previously accepted notions of what is "essentially Southern" has dominated Southern literary studies for the last twenty to thirty years. To understand Lee Smith's early investigation of identity in the new sociopolitical climate—particularly her interest in the self's reaction to perceived liberation/threat—it is helpful to review the deconstructions of traditional definitions of the South as well as the broader postmodern critique of "essential" identity.

Two of the most provocative books written on the subject in the last fifteen years are Michael Kreyling's *Inventing Southern Literature* (1998) and Patricia Yaeger's *Dirt and Desire: Reconstructing Southern Women's Writing, 1930-1990* (2000). Both books are bent on freeing Southern literature from what the authors argue is a stultifying perspective on the South and its culture. Challenging the view of such icons as Louis Rubin and Cleanth Brooks that "Southern literature [provided]...an untroubled rendition of the 'facts' of southern life" (Kreyling xii), Kreyling aims to unveil the shortcomings of this view and expose its complicity in a historical "amnesia" (xii) that enables and perpetuates violence and racism. He bases his sociopolitical critique on the assertion that interpretation of a work is "contingent upon a working literary-cultural process of identity" (x). Similarly, but more radically, Yaeger focuses her own study on the "throwaways" of the South (xii), people of the region whose experience and value cannot be understood in the framework of traditional categories. Her goal is "to make the usual expectations [of Southern literature] strange—to explore the density and peculiarity of Southern women's fictions across racial boundaries" (ix). Like Kreyling, she is plagued by the violence and degradation of life beneath the façade and by what is *not* heard. But Yaeger goes further to assert that Southern literature is best understood as reflecting "crises of whiteness" (11).

Such frank reconsiderations of Southern literature have rejuvenated it with relevance in an era when most readers and scholars are, like Yaeger, "tired of these categories" (ix), such as the belle and the often narrowly represented Southern family. Such arguments certainly not only open up notions of "Southernness," but by their very nature, critique definitions of identity that collaborate with "history" to invalidate plurality of experience. Both Kreyling and Yaeger attempt to right some of the wrongs done to marginalized groups, African American Southerners in particular (Yaeger addresses gay and lesbian Southerners as well), by "reinventing" and "reconstructing" Southern literature as a more pluralistic, and less Southern Agrarian-defined, body of work.

What I would like to emphasize in discussing these two scholars is their insistence that the old categories do not authentically characterize the South, either historically or presently, nor are they sufficient help to the contemporary Southerner negotiating the complexities of his/her lived experience. Smith is a white Southern writer who deals with race only peripherally—her protagonists are all white except *The Devil's Dream*'s Jake Toney and R.C. Bailey, both of the mysterious Melungeon background. Yet, Kreyling's and Yaeger's arguments illuminate Smith's insistence that the old categories, like those of any culture, function not to unveil truths of the culture but to shape it toward a sentimental view, which for many people is a suffocating "memory" of who they are.

More broadly, Smith's characters, even those of the early novels, struggle with, and move toward embrace of, the postmodern era's "indeterminacies[,]…ambiguities, ruptures, and displacements affecting knowledge and society" (Hassan 504). Unlike Kreyling and Yaeger, they do not have a sense of "the big picture" in which their negotiations play a role; rather, they are just waking to the inadequacies of the roles they have inherited. Their waking is of course liberating, enabling validation of experience not prescribed by the old categories. But during this awakening, Smith's characters also experience the disorientation and uncertainty of travelling without the old maps.

In *The Last Day the Dogbushes Bloomed*, *Something in the Wind*, and *Fancy Strut*, Smith's characters find themselves between a rock—the socially restricted and often suffocating position they are expected to fill—and a hard place—the maplessness of the "post-South." In this context, it is important to note that the developing awareness of the old categories and their inadequacies does not negate the power of those categories. In Jane Flax's *Thinking Fragments* (1990), Flax quotes G.W.F. Hegel to say that "whatever happens, every individual is a child of his time" (3). Exploring the ways that "psychoanalysis, feminist theories, and postmodern philosophies intersect in their common fascinations," Flax emphasizes their efforts to formulate understanding less *conclusive* than *"process-oriented"* (3, my italics). As she begins her investigations, she acknowledges not only the promise of the deconstructive approach—its potential to deal more honestly and justly with issues of pluralism—but also its other edge: In this new space where points on maps are only temporary markers whose meanings are always to be suspect, she asks, "How is it possible to write?...I believe many persons within contemporary Western culture share such feelings of unease, of being without a secure ground or point of reference" (5). Such is the situation of many of Smith's characters.

Smith's early fiction finds her characters in this place of negotiation, often pushing against the boundaries of their cultures' prescribed values, but nonetheless children of their time and place. Traditional ideologies are not less powerful for being inadequate structures. Smith pursues investigation of the *self* in these early novels by employing a key strategy that she goes on to develop with more impact in the more sophisticated novels that follow: portrayal of the *split self* that can occur when one's desires or inclinations cannot be reconciled to society's expectations. One can see, in Smith's early work, her efforts to explore notions of the split self toward the ends described above: to observe characters' chafing against prescribed Southern roles and to posit what a "liberated" self would mean.

Admittedly, I employ the term "split self" quite liberally here. I am no psychologist and neither is Smith. Yet, discussion in the field of psychology, particularly psychoanalysis, about defense mechanisms helps to illuminate Smith's interests. Uwe Hentchel, et. al, provide a thorough overview of these mechanisms in *Defense Mechanisms: Theoretical, Research and Clinical Perspectives* (2004). Acknowledging Freud's establishment of the concept, as well as significant revisions of it that have occurred since, the authors explain that "In the classic psychoanalytic view, defenses are directed against internal danger. Such a danger leads to the experience of intrapsychic conflict" (6) and can include physical threat, guilt, and loss. Further, "it has been demonstrated that defense mechanisms are invoked response to threats to self-esteem, identity status, object self, and core personal beliefs" (6). Dissociation, in which "Conflicts and stress are dealt with by temporary failure of consciousness to integrate the dangerous material" (Kline 47), is one such mechanism.

Dissociation can be manifest on a broad spectrum, at the more functional end causing a sense of "depersonalization," the feeling of "watching oneself" while having no control over a situation, or the sense that one is living "in a dream" (Moskowitz and Evans 198). In more extreme cases, it is associated with a sense of "unreality" or even dissociative identity disorder, previously known as multiple personality disorder (Steinberg 62). Expert perspectives on the conditions and manifestations of dissociation are, of course, greatly varied in the field of psychology, and I do not purport to know which is most accurate. My reason for introducing the concept into this discussion is strictly to provide a framework for understanding Lee Smith's identity-exploration. Throughout her fiction, we see characters experiencing this sense of a "split self," sometimes exhibiting symptoms that look much like the above descriptions of depersonalization and at others experiencing the more serious sense of "unreality" that can occur with dissociation.

In *The Last Day the Dog Bushes Bloomed*, Smith experiments with the first-person narrative, that of nine-year-old Susan Tobey, to explore the taboos of human experience—in this case the violence, both physical and psychological, that can accompany the transition from innocence to knowledge. A factor which complicates Susan's ability to navigate this transition is her sense of the codes of the genteel South, codes which ostensibly govern the behavior of her family and their expectations of her. The story depends upon a rather conventional use of the coming-of-age novel genre and employs a somewhat awkward allegorical tone, yielding a contrived ending in which Susan has, at the age of nine, lost both her virginity and her mother (whom Susan has deemed "the Queen"), who has left home ("the Castle") to be with her lover ("the Baron"). Yet, through Susan's telling of the story, Smith effectively reveals both the joys of Susan's vivid imagination as well as her attraction to and fear of adult knowledge. Perpetuating her premature coming of age are two key factors: 1) her mother is having an affair, triggering the final disintegration of her parents' marriage, and 2) the neighborhood children's natural interest in the sexual is cultivated and brutally exploited by their troubled peer, Eugene, who is in town visiting his grandmother for the summer.

I Have Been So Many People

From the beginning, Smith illustrates Susan's imaginative perspective. The nine-year-old spends a good deal of time alone in her secret place beneath the "dogbushes," or hydrangea, and in the "wading house," a shady spot in the creek nearby, where a willow arches over the watery home of her friends, a lizard named Jerry, a turtle, and a baby black snake (12-13). Smith conveys much through Susan's description of the wading house:

> *It was a soft, light green tree, a willow, that grew by the bank of the stream. The way the branches came down, they made a little house inside of them…I was the only one that knew about it. It was a very special place…A very wise old turtle lived there too. He blinked his eyes slow at me and I could tell that he knew everything there was to know.* (12)

In addition to her imaginative connection to the natural world, Susan feels a strong bond with the family's African American maid, Elsie Mae, who, according to Susan, "was old and hopped around. She had the best feet I ever knew. They were the littlest feet in the whole world" (2). She loves Elsie Mae, whose feet are always "tapping" (14), and her stories, which open up a new world to Susan, often teaching her about life experiences that her parents have not prepared her for. When Susan asks Elsie Mae if she was ever in love and if so, why she did not marry, Elsie Mae replies, "Oh, law…You don't have to marry somebody just because of you love them…There ain't no rhyme or reason to the thing at all. You just can't ever tell" (112). Susan's descriptions of these relationships serve, in the narrative, to reveal a self-identity being negotiated, both nature and Elsie Mae's stories giving her a glimpse into the more complicated world beyond her simple childhood.

Yet, as she relates the challenges to her innocence that summer, she reveals a sense of her consciousness "splitting" at moments of particular duress. For example, upon meeting Eugene, she steps outside her childish unselfconscious state to judge herself harshly. In showing around the strange little boy as her mother has directed her to do, she takes him to see her family's barn, and once there, she is delighted to see a baby mouse sitting next to a post. Eugene's propensity for destruction shows itself at even this early age as he throws a rock at the mouse. Shocked, Susan yells at him, "'Why did you throw that rock?' I said. 'Why?' My voice was just like a baby's and I knew it and I hated it" (22). Susan suddenly sees herself from what she imagines to be Eugene's point of view, and the negative self-judgment is a hint of what will happen on a more extreme level, later in the narrative, when her morals are challenged more seriously.

Eugene, who serves as an agent of the dark for Susan and her friends, introduces them to Little Arthur, his imaginary proxy. Little Arthur, Susan comes to understand, also embodies the "perverse" impulses inside each of them. He thrills them by telling them, by way of Eugene, to mix the blood from their cuts and scrapes and to look at an art book featuring statues of naked figures. Although Susan willingly follows Little Arthur's instructions, the challenge to her previous sense of self threatens her psychological stability. When Susan and the other kids dig up Mrs. Tate's beautiful rose garden at Little Arthur's insistence, she begins to laugh. For Smith, this particular brand of laugher, as we see later in *Something in the Wind*, is a symptom of the self's splitting: "Then we pulled them [the rose bushes] up and threw them on the ground and laughed, and I could feel the house on the hill look at me and I laughed harder…We stopped where we were and my stomach felt funny and sick when my laughter had ended" (159).

This episode climaxes when Susan is raped by Eugene in a strange ritual-like episode of "doctor" in Mrs. Tate's rose garden, and here, her perspective reveals the violence her new knowledge brings, both physically and psychologically:

> *They [Eugene and Robert] jumped on me and pulled down my shorts and they were going up and down, up and down, up and down, on me with no clothes on, and they were the Iron Lung, and the doctor said, "One, two, one two." The nurse said, "I think you're going too fast." I was going down and down into the earth, and the Iron Lung was hurting me between my legs, and the dirt was coming up from all around to cover me, cool and friendly, coming up to cover me because I was dying.* (163)

When Eugene is finished, "The Iron Lung jumped off me and I could see again, but I couldn't get up from the ground. My legs wouldn't work…I looked at my legs and they weren't tan anymore; they looked like white rock in the moonlight" (163). Susan's observation of her body here shows a certain level of dissociation from it: it looks and feels strange to her, and she cannot control it.

This encounter, along with her mother's leaving and the death of the family's yard man Frank, requires Susan to deal with a new understanding, that of sexuality and its close connection with death, which also leaves her with a new self to negotiate. In order to cope psychologically with all that has happened to her, Susan compartmentalizes the events and knowledge:

> *A lot of things reminded me of the Queen but I had a new trick of how not to think about those things. I had fixed my mind up so it was cut into boxes, sort of like the boxes eggs come in. In*

> *one box, I put the Queen; and in one box I put Little Arthur, who was not dead either but alive all the time and I knew it; and in the other boxes I put the things I liked...That way, if I ever wanted to think about anything I could just pull it out of its box and roll it around in the part of my head that was not boxed in. When I got tired of it I could close it back up in its box, and there were some things that I never took out of their boxes at all.* (173)

She ends the story with the implication that she will survive the transition: "I turned around and smiled at Little Arthur to show him that I wasn't afraid of him anymore. Then I went upstairs and put on my new yellow dress and my new red shoes without straps and I went out to dinner with daddy" (180). This ending seems, at least temporarily, to resolve the tension of Susan's coming-of-age challenges. Through her well-developed imagination, she is able to hold together the disparate and threatening factors of her experience. But perhaps more interesting than the resolution here is Smith's portrayal of the self's tendency to temporarily split, illustrating her first attempt to investigate an evolving self and the psychological turmoil it must face to continue functioning in a prescribed social role.

Smith's second novel, *Something in the Wind*, also employs first-person point of view, except Brooke Kinkaid navigates the difficult waters and the resulting split self of early Southern womanhood rather than of adolescence. An older protagonist allows Smith to focus more directly on the search for an identity in context of a stifling traditional culture. Katherine Kearns notes the challenge that Brooke faces in her assertion that "[s]he [Brooke] wants the approval of her family but can attain it only by making certain that her successes fall within the acceptable parameters" (182). Caren J. Town, in *The New Southern Girl*, discusses the commonality of this plot in Southern women's fiction: "Many of the young female characters of the past find themselves feeling lost, hopeless, or at best aimless at the end of their novels, as they fail to fulfill the wishes of their parents, the expectations of their communities, and the demands of their traditions" (18). Brooke desires to please her family and friends but cannot seem to fit their idea of who she should be, and her narrative of her experience reflects the stress of her position. She conveys in her telling of the story the moments of serious dissociation that result. At the funeral of her close friend Charles Hughes, she acts on her impulse to stand a little in the church pew in order to see Charles's body while the sitting congregation prays. Brooke's failure to conform to the service's conventions angers her older sister: "'Brooke, what do you think you're doing?' Liz said, hissing, and I saw that she was mad. You know how to act, Daddy had said, only I obviously did not. I sat back down. 'I'm sorry, I didn't know I was doing that,' I said to Liz, which made it worse than ever" (15). In context of the funeral, so dependent on the formalities of tradition, the ambiguities of Brooke's identity are especially pronounced.

An English major when she was at college, Brooke retrospectively tells the story of her journey from a consciously bifurcated self—the unruly Brooke Kinkaid and "Brooke Proper"—to a self with courage enough to break free from Brooke Proper and see where this decision will take her. Negotiating mental instability along the way, including dissociation, Brooke reflects the extreme consequences that can result from the strict norms of traditional Southern society and their intolerance of nonconformity. About halfway into her freshman year of college, having "played the game" as well as she knows how, Brooke comes to the point of complete dissociation from the reality of the woods she has walked to, dressed in nothing but her coat, to get away from her conventional college boyfriend Houston: "I lay down in the snow to rest for a minute. It was very shallow, as soft as a layer of feathers on the ground. One by one all the people came out of the trees. They began to dance around me, singing a song with garbled words in their high, thin voices" (86). When Houston catches up to and speaks to her, worried that something is wrong, she does not even recognize him at first. The business of "faking" a self that others can accept has taken such a toll on Brooke's psyche that she can no longer tell what is "real" and what is not. Brooke has been participating in a kind of "faking" that perhaps all people must in negotiating a traditional culture as omnipresent as that of the South; it is her uncommon sensitivity to the lack of inauthenticity that pushes her toward a psychological split.

It is her affair with another student, the strange Bentley T. Hooks, that finally enables her to explore the dark and unfamiliar depths of her mind. At college on a golf scholarship, Bentley comes from a poor, evangelical family, featured himself as a prophetic visionary in revivals when he was a child. His behavior is outside of polite society's mores—upon meeting Brooke on campus, he asks her directly to have sex with him, and throughout the novel he is plagued by the "visions" with which he has been "gifted" since his early life. Brooke's decision to move into "the pit" (Bentley's basement apartment) seems a step in the right direction for her. She abandons her game of imitating her roommates, Diana, the Tri Delt who always knew "which end was up" (52), and Elizabeth, who according to Brooke, "had a still point inside her and everything spun out from the center" (112). When Brooke is with Bentley, she lets down her mask, and this seems to bring her closer to mental health, to a sense that her "unfixed" self is more authentic than the other identities she has been trying on: "From the time that we packed all my gear in the Volkswagen and started off for the pit, I felt real. Everything that happened to us was really happening" (154).

This relationship has its own problems; Bentley's own mental instabilities and violent tendencies ultimately motivate Brooke to move out of "the pit." Yet, the experience has liberated her to some extent. It has given her the chance to learn about her authentic self in a space uncontrolled by the mores of her family and her more conventional college friends.

Brooke ends her narrative with her brother Carter's wedding, in which she is bridesmaid:

> *I could have this too, I thought. I could marry John Howard and step into my place like Carter and have all this for the rest of my life. But I didn't think I would. I had come full circle myself, and now there were new directions. Speaking of which, I decided I would call Elizabeth and see if I could go to Nova Scotia with her and her family in July. She had asked me before she left school. That meant Carolyn would have to go to Gloucester by herself. She would just die; I could hear her now. I started laughing.* (214)

Brooke's laughter, at the close of the story, leaves us wondering if she is still somewhat unstable—her laughter often bubbles up when the restrictions of her culture seem to threaten the possibility of authentic experience. Smith does not indicate that Brooke has found a space in which the "real" Brooke Kinkaid can live unassaulted. But she does suggest, in Brooke's retrospective narrative of these years, that she has survived and has even been able to find meaning in those troubled times.

In her third novel, *Fancy Strut*, Smith experiments with multiple narrative perspectives in her quest to examine identity on a broader scale. A product of the still-developing writer, this novel has been acknowledged mainly for its comic appeal. Lucinda MacKethan comments that Smith registers, in *Fancy Strut*, "the comic potential in small-town talk, hot local love affairs, aspirations, and machinations of all sorts" ("Artists and Beauticians" 7). Similarly, Kearns praises this "broadly allegorical comedy" (176) as a technical/artistic success: "The display of pyrotechnics within is Smith's own fancy strut across the page" (177). More importantly, however, in this novel Smith explores the identity both of individual characters and of a small Alabama town, through a carefully selected combination of perspectives. Through multiple points of view, the sesquicentennial celebration of 1960s Speed, Alabama, is relayed in all its glory and difficulties, the story shifting strategically from perspective to perspective to convey the complex character of the town's evolution.

The novel opens in the company of the unchanging Miss Iona Flowers, who is as old as the town's newspaper, the *Messenger*. Miss Iona fixates, as "society and ladies' editor of the weekly *Messenger*" (4), on her own idealized version of the events in her small Southern town: "Sitting behind her father's old desk, Miss Iona nibbled delicately at a slice of candied orange peel and mused upon her destiny. She saw herself as the custodian of beauty and truth in Speed, the champion of the pure and good" (4). Daughter of the *Messenger*'s founder, Miss Iona represents the "old guard" of Southern culture, and as such, Smith reveals, Iona creates this version of the South in her "journalism." Ostensibly dedicated to "writing up" the social events of Speed, such as "weddings and anniversaries and parties and club meetings" (4), Iona is known to *generate* many of the details of these occasions as she writes the stories, having placed "Grecian urns of bougainvillea in every home and at every wedding" (4) during one period of her writing.

In relief against Miss Iona's version of the South, then, young Bevo Cartwright provides a perspective that reveals the difficulty of establishing a sense of agency in this context. A timid fifteen-year-old when the novel begins, Bevo is irrationally afraid of fire, and just as notable, he spends his spare time looking for himself in the photographs collected in his mother's countless scrapbooks. Lately, this activity has affected him strangely: "As long as Bevo kept his eyes open and looked at the picture album he was all right, but as soon as he closed them, he felt funny" (90). Bevo worries that without the pictures as a reference, he does not exist at all. In Bevo, Smith portrays the self's splitting in such a way as to render the individual powerless. Unlike Miss Iona, Bevo is not bolstered by the sense of tradition, by the sense that the sentimental ideal will carry him through his developmental challenges. On the contrary, "Bevo felt a terrible obligation to be picturesque" (55).

After getting high for the first time with his friend J.T., however, Bevo becomes bold. He destroys his mother's ever-present Instamatic camera, smashing it with one of his Mamaw's garden tools. From this moment on, he begins to recognize his pent-up frustrations, anger, and desire: "He had never, he suddenly realized, said a cross word aloud to anyone in his life. Wasn't it about time?" (291) In context of his somewhat caricatured family, Bevo comes across as an authentic boy coming of age, beginning to realize that he exists beyond the hundreds of photos his mother has snapped, beyond the "family history." Now, Bevo begins to think about fire in a new way as a result: Rather than fearing it, he becomes fascinated with it. As the bugler at the Sesquicentennial finale, he starts a fire beneath the stage with the matches he has stowed in his Confederate uniform pocket. After the flames have consumed the stage, and the surprised crowd has run away screaming and then come back again to watch, Bevo feels good: "He could take on anything. Inside himself he felt all new; his mind ran free inside his head; and throughout his whole body, he had come into possession" (340).

Bevo's evolution from a little boy afraid to deviate from what he perceives as the family narrative, through these moments of uncertainty about the actual existence of his authentic self, and finally into a young man willing to take risks in order to find agency over his experience, parallels his town's "coming of age," as well.

Pitted against Miss Iona's ideal of Speed and its history, Lloyd Warner's perspective in the narrative reveals the ugly underbelly of the old categories. A lawyer from an old Speed family, Lloyd is perhaps more burdened by the town's history than any other character. Although he left Speed as a young man, falling in love during his residence in New York, he was drawn back to the small Southern town, much to his chagrin:

> *Only a fool would come back to Speed after so many years. He felt at home here, that was all. Faulkner shit. Then Lloyd was amused by his own bloody little ego and sad because he was amused and because in the end that's what it always came back to. He was too goddamn smart. Always had been, and knew it, and knew he knew it, which made it worse. It was unbearable the things he knew about himself.* (121)

Trapped in his own self-consciousness and his consciousness of the history he carries on his back, Lloyd pretends to be a drunk to prevent the townspeople from attributing too much weight to his words and actions. Then, in context of the Civil Rights Movement, which posed a difficult challenge to many small Southern towns like Speed, Lloyd nervously agrees to represent an African American college student who has been denied an apartment in town by a racist landlord. In spite of the comic tone of the novel, it is clear that Lloyd cares a great deal about righting this wrong, one small manifestation of the greater wrong that was slavery. Even in the midst of this important case, he is suicidal and washed-out after years of trying to reconcile himself to his past. Surprisingly, however, the events of the sesquicentennial celebration, which ultimately spin out of control, provide an unexpected resolution for Lloyd. When he shoots himself in the shoulder during the reenactment of a Civil War battle, having meant to kill himself during the chaos of the scene, he realizes that 'Now he would not have to do it again, was free of it, was free" (328).

While it may seem odd to assert that Speed experiences a "split self" as the story nears its end, this view is a fruitful way of understanding what happens to the small town. Not only has Bevo begun to develop an awareness that he exists as a complex young man beyond the crafted "family history" forwarded by his mother's scrapbooks, but the town as a whole has been confronted by the discrepancy between their sesquicentennial "story," composed by the White Company in a carefully choreographed pageant, and the burgeoning Civil Rights Movement, represented in Speed by Theolester and his African American college peers. Leading up to the climactic stage fire, the town has reached a crisis of identity. But, when Bevo sets fire to the stage during the pageant's finale, he and many others feel released from the stranglehold of Speed's "history," preserved by people like Miss Iona. Although Miss Iona brings the novel to its conclusion, having found a certain way to "slit" her eyes so that her idealist view of Speed and of life remains untouched, Speed's pageant has been destroyed by the fire, perhaps freeing it, on some level, of its old identity and bringing it into the contemporary era.

I have used the loosely defined concept of a "split self" in this chapter to illustrate Smith's early interest in the unique position of the Southerner in the postmodern era. Susan, Brooke, and Bevo, among other characters of these first three novels, face the pressures inherent in the "old categories," the problem of understanding oneself through a "history" that tells only part of the story, and which idealizes the past in the process. Though Smith is still in development as a fiction writer in these works, they show her experimenting with ways to examine the positions of her subjects. The crises these characters face result from their being caught between dissatisfaction with what is expected and the fear and disorientation of acting outside those expectations. In later works, Smith's narratives are played out with more sophistication—sometimes resisting resolution and sometimes offering very credible artistic possibilities for addressing the disorientation. Nonetheless, these early novels reveal the artist's fascination with the challenges of postmodern identity and agency.

Chapter Two
The Drowning of Crystal Spangler in Black Mountain Breakdown

> *She hates to leave the river. Beside it in the dark, she can think it is like her daddy told her it used to be, not flat and dried out and little, but big and wide and full of water. The Levisa River. With huge log rafts on it floating down through the mountains in spring and early summer to the sawmills in Catlettsburg, Kentucky. Sometimes men rode those logs all the way, Daddy said. In the 1920's. Just sitting and floating, it would take days, watching the land coming at you on either side like a dream, the green trees hanging into the water, not ever knowing what would be around a bend...And the water would be clear with fish in it. You could see straight to the bottom. Now the water is black because they wash coal in it upriver...The real black rock, the one Daddy said they named the town for, doesn't even exist anymore...[W]hen the Norfolk and Western came through in the thirties and built the railroad, they blasted the rock into little bitty pieces and it fell into the river and was gone. Probably you could find a piece of it now, in the river by Hoot Owl, if you knew where to look. (*Black Mountain Breakdown* 4)*

Lee Smith's *Black Mountain Breakdown* is often considered her darkest and most tragic novel. This novel's protagonist, Crystal Spangler, "blond and fair, with features so fine they don't look real sometimes" (4), seems to fall victim to the desires of all those around her. Rather than fulfilling the dreams of her own heart, or even recognizing them, she fulfills the dreams of others and, unlike *Something in the Wind*'s Brooke Kinkaid, becomes unable to act as an agent in the world at all, lacking "something, some hard thing inside her" (8), that might endure the assaults upon her *self* that life levels at her.

In understanding the conditions by which Crystal's identity is obliterated, the above passage is tell-tale. The literal black mountain breakdown, the breaking apart of a physical rock whose name supplied the identity for the place, illustrates a constant problem for Crystal's friends and family, nay, for humans at large, perhaps especially in the postmodern era. This problem is the inability to attain, or even to glimpse, the much-longed-for essence of things. In describing Crystal's experience of the Levisa River, Smith's third-person narrator emphasizes Crystal's imagined version of the river, one very unlike the river she sees, "flat and dried out and little" (4), black rather than clear. The entire passage reflects Crystal's fashioned "reality," her relationship to a past and a place that she experiences "like a dream" (4) rather than in its physical form. This is a crucial theme of the novel—that humans are, paradoxically, forever destroying what they love by trying to name the unknown, by attempting to reduce and demystify the abyss, which includes the self. As is discussed in this study's introduction, Southern culture is in a constant state of "being named," by Southerners and non-Southerners alike. Yet, as Patricia Yaeger notes, this naming has usually reduced what is Southern to stultifying types and a shallow understanding of its actual rich and multifaceted legacy. As she does here, Smith will explore in later works the power of art as a way to sustain an authentic self, even one under assault. However, *Black Mountain Breakdown* generates a comparatively dark tone regarding the subject because Smith offers no solution here. Still, in this work, her examination of the condition is so hauntingly beautiful that it leaves a lasting impression which resonates long after we put the novel down.

Critics have often noted the importance of Crystal's objectification in this story, as a young woman coming of age in the Southern, patriarchal culture of Appalachian Black Rock, Virginia. Minrose Gwin states, "For the southern daughter in the patriarchal house, place and identity become compounded and conflicted because place/identity equal(s) powerlessness" (419). Also emphasizing the connection between Southern expectations of women and Crystal's psychological destruction, Harriette C. Buchanan asserts that

> *Crystal's failure to find the grace of laughter or to act with stoic acceptance, as her mother does, is the source of her tragedy. In talking about this novel, Smith describes one of her motives for writing the story: "to really make a thematic point...that if you're entirely a passive person, you're going to get in big trouble. The way so many women, and I think particularly Southern women, are raised to make themselves fit the image*

that other people set out for them, and that was Crystal's great tragedy, that she wasn't able to get her own self-definition. (331)

Further expanding upon the notion that Crystal's destruction is a gender-based tragedy, Anne Goodwyn Jones argues,

Smith is exact and devastating in her portrayal of women, usually married, caught in a cycle of guilt, self-deprecation, entrapment, rebellion, and again guilt that screens them from themselves. The men in their lives...can neither comprehend nor help the woman; most often they contribute to her problem by imposing a fantasy screen of womanhood and refusing or failing to hear the messages from the interior. ("The World of Lee Smith" 253)

Certainly, Smith has addressed in her body of fiction the experience of the Southern woman, often the Appalachian woman specifically, and the limits placed on her as a result of her location in place, culture, and time. At the same time, she has never failed to explore the richness of that very folk culture and the loss incurred as it is gradually displaced by a more modern, progressive yet materialistic mass culture. Further, while gender and culture are factors always at the forefront of Smith's fiction, her work also reveals a fascination with the human mind and its perceptions of itself and the world around it. If we focus only on Smith's exploration of female repression, we are in danger of missing her insightful examination of the human tendency to write the world around us, our tendency to fill the mysterious void with the imagined, specifically with that which confirms our own ideologies.

For many of the novel's characters, Crystal becomes the primary site of this process. Her mother Lorene is constantly planning who Crystal will be. And these plans necessarily depend on who Lorene needs to be. "I may not be a lady," Lorene says, "but by God I'll dress like one" (13). Early in her life, when she realizes that her marital life will not fulfill her expectations, Lorene "center[s] herself firmly in the child [Crystal]" (14). It is important for Lorene to see herself as tough, as someone who does not give in to adversity. Yet, she focuses on Crystal as a reflection of her success as a mother: "Crystal will grow up to be somebody; Lorene will see to that. Crystal will go to a fine school on that rivet money. She will marry a doctor. But whatever she does, she will be somebody special, because Lorene is raising her that way" (14). Crystal seems to recognize the trap on some level, though she cannot fully escape it. She perhaps identifies with her baby cousin in that "[h]er aunt Susie keeps Denny so dressed up that Crystal has never really seen what he looks like" (38). Just as Susie dresses up Denny to look like the baby she wants him to be, Lorene imposes her ideals on Crystal, obscuring any bits of her daughter that might not fit those ideals. Conversely, Crystal's best friend Agnes needs to believe that Crystal is doomed to failure. According to Lucinda MacKethan, Agnes is "Crystal's foil, responding with proper envy or embarrassment or indignation to Crystal's deviations from the norms of Black Rock" ("Artists and Beauticians" 8). In order to hold onto the belief that her own body and personality are acceptable, she must believe that Crystal is shallow and self-destructive: "Maybe if Crystal hadn't been born looking like that, Agnes thinks, maybe that was the trouble all along. Crystal's famous beauty" (151). Although Lorene is determined that her daughter will be "someone" while Agnes maintains her notion of Crystal as a reckless and wayward soul, both do the same kind of damage to Crystal herself. By imagining her as the figure they need her to be, they obscure and perhaps help suffocate the Crystal that lives repressed within her.

Crystal's father plays an interesting role in this process. Grant Spangler is one of the few people who does not seem bent on molding her into a particular shape. But his effect on her exaggerates her inability to grasp a sense of self. Having himself become mostly absent from his family's and town's life, he models a withdrawal that Crystal will later imitate in her final self-paralysis. Unable to pursue his true desires in everyday Black Rock, he begins to live only in the closed-off, darkened front room with his books. He is a romantic and chooses his romantic vision of human experience and of the past over the life he would see if he looked out his window. Crystal clings to this world, as well, through her father. Anne Goodwyn Jones comments on the romantic bond that results from these dynamics: "Crystal loves her father with a romantic intensity that he encourages with his *One Hundred and One Famous Poems*" (262). Crystal loves her father's stories, but she is put off by those that depict disturbing events, like "I Have a Rendez-vous with Death" and "Little Boy Blue," and begs instead for "the daffodils" and "the spider and the fly" (16-17) Ultimately, even though Crystal has had these years with her father in a world where her imagination could be fired, this relationship has not offered her a way to form a stable sense of self; it is too far removed from the world outside her father's room.

Finally, and perhaps most destructively, Roger Lee Combs, whose image of himself takes shape, to some extent, during his first relationship with Crystal, clings to his complementary image of Crystal with a dangerous grip. Having become utterly devoted to Crystal during their time as high school sweethearts, Roger Lee is devastated when she breaks up with him and begins to see Mack Stiltner, a boy who exists on the fringes of high school society. When Crystal returns to Black Rock after her years in college, her affair with Jerold Kukafka, and his suicide, Roger Lee determines that now he will have her. Ironically, this short time back in Black Rock has given Crystal the first chance to establish a sense of personal strength that she has ever had. She has become an English teacher and is surprised to find that she is good with her students. Yet, she is still not prepared to resist Roger Lee's will. Certainly, this scene is brutal, Roger Lee taking possession of Crystal's will, saying, "*Baby.* I've known you all your life, remember? I know you, sweetheart, I know everything you've been and everything you've done...You might *think* you're happy now, Crystal, but you're not" (209). This scene is, as John D. Kalb suggests, a moment of verbal violence as Roger Lee insists that Crystal does not know her self, but that she is what he imagines her to be. Kalb states that by the point of Roger Lee's arrival, "Denial of self has become second nature to Crystal" (25). He argues that the moment of Roger Lee's "victory" over Crystal, signified in his final attainment of her consent to come with him, is a kind of psychological rape paralleling Crystal's earlier physical rape by her mentally handicapped uncle, Devere. Kalb asserts that the moonfaced man in the mental hospital at the end of the novel is not the cause of Crystal's final breakdown, but that instead he is a trigger to a past she has kept buried from her conscious mind, a past containing her rape by Devere, her father's death, Jerold's suicide, and, I would add, a great deal of pain caused by her inability to reconcile her authentic self to the selves she has taken up over the years.

 Yet, lest we easily fall back into our interpretation of Crystal as simply a victim, due specifically to her role as Southern and female, Smith forces us to see that it is more complex than this. Like Crystal's academic older brother Jules, who "knows and has always known too much, has seen both sides of every coin" (82), we are forced to examine carefully the other side of this coin. We are privy to the irony of the narrative because, as MacKethan points out, "the coupling of third-person point of view and present verb tense in *Black Mountain Breakdown* achieves a tension between closeness and distance crucial to both mystery and meaning in the novel" ("Artists and Beauticians" 11). This tension can be seen in the following passage in which Crystal reflects on her new life as Roger's wife, a life which, according to conventional wisdom, should make any woman happy: "Crystal fluffs up the quilted pillow by her side in the white armchair, aimlessly, then smoothes it out on her lap. Fan pattern: yellow and red, picking up the warm dusty-rose shade of the walls. This is a beautiful room. Everybody says so" (217). This passage illustrates the ironic effect of Smith's third-person narrative. While we get neither the narrator's nor Crystal's direct evaluation of Crystal's lack of feeling regarding her "beautiful room," Smith employs short, almost abrupt sentences which voice the judgments of others: "This is a beautiful room. Everybody says so." These sentences appear to reflect Crystal's thoughts. Yet, the objective tone reveals that they are only mechanically repeated thought, not Crystal's own feelings about her situation. This choice of narrative strategy creates an organic reflection of both the process by which others project onto Crystal and Crystal's own psychological tension. Not only is there something peculiar to Crystal that makes her the likely site of even her fellow Southern women's projected meaning, but Crystal herself engages in this continuous process to establish her identity by infusing the exterior void with explanations/stories that can provide satisfactory meaning to her interior self as she understands it.

It is Crystal's particular mysteriousness that makes her the canvas on which others commonly cast their imaginings. For example, though she is Miss Best All-Around and Miss Best Personality at Black Rock High School, the narrator explains that "her eyes are too large and too blue and too deep. When she's not talking to anybody, when she's staring out a window or not listening to a teacher lecture in class, her eyes seem like lakes, as if there are secrets in them, as if a mystery is there" (91). Similarly, Mack Stiltner finds frustrating his relationship with her: "There's a lot of things it's hard to know for sure about Crystal. Mack has been dating her off and on now for about a year, over a year now, and he still can't figure her out" (96). The inability to know Crystal generates, for Mack as for others, both attraction and anxiety.

Even Crystal has trouble getting a fix on herself, and mirrors are no help to her. Mirrors are a recurring motif in Smith's novels, symbolic of the complicated relationship between a person and his or her reflected self. The reflected self becomes, in these moments of confrontation with one's image, a foreign object to be reconciled possibly with one's essential self, but more clearly with one's ideologically defined self. One's self is, in this instance, the void to be dealt with. For example, Ivy Rowe, protagonist of *Fair and Tender Ladies*, is shocked, upon seeing her reflection in a store window, to find that she is beautiful: "Now who is that? I thoght to myself as we turned the corner and waited in front of the pharmacy to cross the street. And...it was us! Us in the winder looking like movie stars, me too. It was such a surprise I like to have got run over crossing the street" (70). Like Ivy, Crystal is separate from this reflected self, even feeling at odds with it at times. When she goes up to Dry Fork to spend time at the family homeplace with her great-aunts, Crystal gets up during the night and steps in front of the dresser: "looking into the wavy, tilted mirror. She sees herself in shadow, backlighted...Who is it there in the mirror? She sees long bright hair and no face, no eyes, no nose, no mouth. Moonlight spreads over the quilt. Who? She wonders, shaping the word with the mouth she doesn't have. *Who?*" (36-7). So early in the narrative, this moment of disorientation foreshadows her lack of psychological resource for dealing with the difficult events ahead.

Later, as she readies herself for the Black Rock High School Beauty Contest, she has a similar moment: "Crystal is perplexed by her made-up face in the mirror. It doesn't go with her hair. Or her hair doesn't fit the face. Anyway, she doesn't look like herself in the mirror" (106). This discrepancy is disturbing to Crystal, this confrontation with a version of herself that does not seem to fit. Yet when Crystal cannot see herself reflected, she often has no sense of self at all. She dates almost any boy who asks, enjoying the image of herself reflected back to her by him: "It's a funny thing, but she doesn't feel real when she's by herself" (39).

Later, when she is married to Roger Lee Combs, who coveted her all their adult life before finally "winning" her, Crystal feels precarious. She can no longer sleep, and to busy her mind, "She makes up other selves," feeling, she thinks, "like a person in a play" (221). It is clear that throughout the novel, Crystal lacks a real sense of self, the kind possessed by survivors like her friend Agnes and her mother Lorene--possessed, significantly, at the expense of Crystal herself.

Yet her objectification as the site for constructing meaning does not preclude her own guilt of the same crime, the same desire to discover meaning she can cling to and a projection of her own desires onto "the void." In this vein, Crystal seeks to understand her world by imagining herself in relationship with the past, that past being an unfixed mystery much like Crystal herself. She escapes into her imaginary games, like that of her fantasy ghost, Clarence B. Oliver, infusing what makes sense to her into a space that is otherwise confusing and scary. As a child, afraid of potentially harmful yellow ghosts and green ghosts, Crystal constructs the protective Clarence B. Oliver to balance the threat. "If you touch the wood and are obedient and fair with the colored ghosts, then Clarence B. Oliver will be there when you need him to help you out" (36). Similarly, as she grows older, she often escapes into sexual or spiritual ecstasy. Having won the Black Rock High School Beauty Contest, she is left feeling empty and alone. To fill this emptiness, a result not only of her loneliness but also of her lack of self-knowledge, she welcomes sexual encounters that help distract her from the inner void:

> *She lets...boys touch her and she doesn't care. Sometimes she lets them go all the way too and she doesn't care about that either...Because it's only when she's with boys that she feels pretty, or popular, or fun. In the way they talk to her and act around her, Crystal can see what they think of her, and then that's the way she is.* (140)

For Crystal, the appeal of attending a tent revival with her young neighbor Jubal Thacker is similar to that of sex. At the climax of the meeting, "[a] current arcs through her body, making her feel like she felt when she was with Mack—alive, fully alive and fully real, more than real" (126). These coping mechanisms obscure, for Crystal as well as for those around her, the terrible truth that she has no sense of an authentic self, and thus no real agency in the world.

As the epigraph suggests, Crystal finds herself unable to reconcile observations of her culture—making the transition from traditional and land-based to capitalistic and media-based--into a satisfactory context for her own identity. She is unable or unwilling to assemble into even a tentative "collage" what she knows about her environment—that the creek is now withering and black, that her father's family is not clinging to its history, as Crystal's hoarding of the ancestral diary would indicate, but is letting go to accommodate the future, as is revealed by the transformation of her stepfather, Odell. Half-brother of Crystal's deceased father, Odell was at one time a mechanic and handy man, awkward and unkempt; he spent most of his time working with his hands and was most comfortable doing so. But his marriage to Lorene turns him around and makes them both happy and settled. Now he drives a luxury car rather than his old pickup truck and wears "a gold-colored polyester jumpsuit" (162) instead of his work clothes. When Crystal tries to get him to remember what kind of beer he always used to drink, insisting that it was Pabst and not Schlitz, Odell answers, "Well, I'll tell you...That might be so. But things change, and you have to just kind of go with it, if you know what I mean" (169). Crystal does not like this change. She is searching for something that has remained the same, something stable. It is difficult, if not impossible, for Crystal to cope with or understand the real world, a world whose essence is elusive. When she eavesdrops on a conversation between her mother and her aunt Neva about a violent conflict between Mrs. Belle Drury and her husband, "In her mind, Crystal has a clear picture of the girls on the covers of these [Gothic] novels (in despair, wearing long pastel satin dresses, fleeing across some dark landscape from the gloomy castle in the background) but no clear image at all of Mrs. Belle Drury who is living through such a hard time" (133). She works hard to pull together a unified idea of the world, by trying to link with her past. When she visits her great-aunts at Dry Fork one night, she goes into the basement poking around and stumbles onto the "old leather-covered journal" (201) of Emma Turlington Field: "This journal—or something like it—is what she has been looking for. Something to establish the past, continuity" (202). She experiences a longing for something solid on which to get a foothold, as is revealed in her memory of her lover Jerold: "These blank spots used to give her a thrill: all the dark unknowns about Jerold, but sometimes she used to wish he had a real past, anything to put your finger on" (188). But in the end, her efforts to reconcile her real past with her imagined past fail. Her marriage to Roger Lee is not enough to establish for Crystal a sense of self, nor does it keep her past pain from coming back to haunt her. After her sexual encounter with the "moonfaced" man in the mental institution, during which she relives her rape by her mentally handicapped Uncle Devere, she is overwhelmed by the past. Later, she sits on the couch in her beautiful home as Roger Lee quizzes her, trying to understand what happened. When she ignores his questions in favor of reading the old leather-bound journal, "Roger comes to her, takes the journal from her lap and closes it and puts it in the fire. 'I want that,' Crystal says, but she makes no move to get it. 'No,' Roger says" (233). Unable to piece together a stable sense of who she is, what the past is, Crystal finally chooses catatonia.

As anxious residents of the postmodern world, we are continuously falling into this trap of trying to write meaning into the void, suggests Smith. Crystal's visit to the modernized Natural Bridge exemplifies this tendency. Crystal has always wanted to stop there, but when she finally gets the chance, it has been transformed into something that is anything but natural: "a loud symphonic recording of 'How Great Thou Art' comes from some mysterious wooden source while they view the bridge" (138). This familiar attempt to shape our experience of a thing by labeling it and imposing meaning on it reveals a particular unease with the transitory nature of reality and its unfixed meanings.

In spite of Crystal's trespasses against the integrity of the void, her peculiar unfixed and undefined quality seems to indicate that she understands on some level the falseness of naming the void. Crystal is subconsciously aware that humans have imposed names on things to understand them and that this act is to some extent artificial and reductive. In imagining what Slate Creek must have looked like when the area was still "wild," she admits to herself, "[o]f course it would not have been named Slate Creek then…There was no coal and so there would have been no slate either, just a big creek without a name and a hollow tree there, cut out by lightning perhaps" (9). Often, Crystal seems drawn to a state where she can exist as indefinable. She enjoys going up to the holler where Mack lives because "[n]othing has a name up here" (98). And when she becomes fascinated by the brand of salvation offered by Jubal Thacker, she feels she has found an ideal state, wherein "Crystal is nothing but flame" (127). She likes "being nothing at all" (128).

    Yet in this state, Crystal has no sense of a personal center. As Jane Flax warns, while the decentered self is attractive for its tendency to liberate the mind, a complete lack of center can leave one without the ability even to function (219). In this extreme state of fluidity, Crystal drowns. This drowning occurs in contrast to her mother Lorene, who, coming up from a reverie of what her life could have been like, "shakes her head slowly and decisively back and forth, like a swimmer coming up from under the water. It's not noon yet. She has things to do. And anyway, she's not through with Crystal; Crystal is still at home" (25). MacKethan notes, "Lorene and Agnes are people who make life work within recognized limitations and without losing their own special poetry of spirit: 'Lorene always heralds the approach of any new season as if it were a person'" ("Artists and Beauticians" 9). While Lorene survives by vigorously taking up the role of her ideologically defined self, Crystal refuses to impose that kind of identity on others and cannot adopt one for herself. The result is ultimately her complete retreat from the world, with Agnes sitting beside her bed, stroking her arms, believing that she is happier this way. Even in her catatonic state, the mystery of Crystal is still keeping others afloat. And this is the dark ending of the novel. We are unsure who is to blame, who is the villain in this story. Yet those of us who will read on in Smith's body of work will find that finally art, more importantly creative acts, will provide a more robust protean identity that might be taken up--an identity that, while not fixed, offers substance and agency. With this insight, as we reflect back on *Black Mountain Breakdown*, we can see that the novel, the third-person narrative itself, offers a kind of respectful impression of who Crystal is, both mysterious and carefully rendered.

Chapter Three
Narrative Mourning: Textual Suspension of Past/Present in Oral History

She pulled something little out of the pocket, pulled it out slow and painful, the way she does everything, and then she let out the awfulest low sad wail I ever heard. It did not sound like a person at all. It sounded like something right out of the burying ground, something rising up of age and pain. (*Oral History* 277)

At the climax of Lee Smith's *Oral History* (1983), Sally Cantrell tells of Ora Mae's pulling from her pocket a pair of golden earrings, handed down from Pricey Jane Cantrell through the line of women that followed her, and throwing them over the cliff. This ceremonial moment seems to be a turning point in the novel since Ora Mae's action, so long in coming, ostensibly breaks the curse cast on the Cantrells by the witch, Red Emmy, so long ago. But is the curse broken? Do the Cantrells stop suffering from desire? And more importantly, Smith asks, do we, the readers, suffer from the same curse, mourning for what we desire but cannot have? Through the apt metaphor of the golden earrings, Smith leads us, once again, to explore the problem of an unknowable reality, which includes the unknowable self.

In this cleverly crafted narrative, Smith reveals that we, her readers, are not so unlike the Cantrell family, "haunted...every one of them all eat up with wanting something they haven't got" (235). Like Almarine Cantrell, who, after the death of his wife Pricey Jane, harbors "a sense of the void which opened up when Harve nailed Pricey Jane into the box" (93), the reader is held to this novel by her or his own sense of the void, the absence of the thing that is the past. By presenting this written, fictional "oral history," Smith draws our attention to the paradox in both the narrative and in our own yearning for a past, and hence a history-based cultural identity, that in the most literal sense does not exist.

Smith's *Oral History* provides a fertile context for the discussion raised by Fred Hobson in *The Southern Writer in the Postmodern World*. Hobson argues that the contemporary phase of Southern literature is distinct from "classic" Southern literature (exemplified by the likes of Faulkner and O'Connor) in that unlike their predecessors, Lee Smith, Bobbie Ann Mason, Clyde Edgerton, and other writers of the late twentieth century "immerse their characters in a world of popular or mass culture, and their characters' perceptions of place, family, community, and even myth are greatly conditioned by popular culture..." (10). Hobson rightly asserts that much of contemporary Southern literature exhibits a move away from earlier Southern "novels of ideas, the novel of historical meditation, or the novel concerned with sweeping social change" (10). However, he sees in the current phase "a relative want of power, a power that often had its origins in or at least was related to—in Warren, Styron, and part of Faulkner—a certain southern self-consciousness" (10). While Hobson's observations about the media-influenced environment of the postmodern South are generally accurate, his conclusion that the New South is no longer struggling with its past, and thus that novels by writers like Smith do not generate the impact of their Southern predecessors, is somewhat shortsighted. Smith and her contemporaries are still exploring questions of identity, but in a world further complicated by the understanding that "history" as it is presented to us is mediated, nay, even constructed. The guilt of slavery still hovers over the South like a ghost, but the "memory" is made more complex by the "packaging" of Southern history in a number of versions. And this history is juxtaposed against the more homogenized human culture being shaped by media and capitalism. *Oral History*, and the oeuvre of Smith's work as a whole, reflects the contemporary problem of identity, stemming in part from the psychological dependence on a past that we have, to a great extent, imagined. Recognizing the elusiveness of that past, we experience an anxiety regarding our identities, which can seem rootless and unstable.

Smith reveals, in the story of the Cantrells, a desire we recognize, a yearning that can never be authentically alleviated. Just as for the contemporary reader, the identity of the Cantrells is in a kind of limbo, between what they have and what they desire. This limbo can be observed in Dory, who yearns for her lover Richard Burlage and the life he might have given her. Out of his own desire for Dory, Little Luther Wade composes a song that captures her longing for Burlage:

> *Darlin' Dory stands by the cabin door*
> *Standing with her Bible in her hands*

I Have Been So Many People

> *Darlin' Dory stands by the cabin door*
> *A-pinin' for her city man.*
>
> *You can throw that Bible down on the floor*
> *You can throw it out in the rain*
> *Prayin' for him all night long won't do no good*
> *For he ain't a-comin' back again.* (172)

Notably, Dory's longings are voiced not by her, but by others. Harriette C. Buchanan describes Dory as "[t]he passive heroine, yearning for some undefined and perhaps indefinable something that will deliver her into a life somehow more or better" (334). Like the past we yearn for, and like the love Dory pines for, Dory herself (like *Black Mountain Breakdown's* Crystal Spangler) is, for the reader and for those who know her, a sought-after mystery. Buchanan notes, "Dory is the center, but her reality remains a mystery because her story is told by first-person narrators [other than herself]" (338). This limbo is an apt illustration of the condition generated by the human desire to know our reality, including our past, and our ultimate inability to do so. The limbo created by desire is also reflected in Almarine Jr. and his family, who by the end of the novel have agreed to sell off part of the long-standing family homeplace for the purpose of building the theme park, Ghostland.

> *Al...will make a killing in AmWay and retire from it young, sinking his money in land. He will...embark on his grandest plan yet: Ghostland, the wildly successful theme park and recreation area (campground, motel, Olympic-size pool, waterslide and gift shop) in Hoot Owl Holler...And the old homeplace still stands, smack in the middle of Ghostland, untouched...[E]very summer night at sunset...those who have paid the extra $4.50 to be here...sit in this cool misty hush.* (285)

This deal will ostensibly bring the Cantrell family closer to the level of material prosperity that Al assumes will make them happier, the kind measured in plush vans and bass boats. Although the reader recognizes that the material luxury Al seeks is essentially an illusion promoted by advertising, it is Al's desire that to a great extent defines him. The reader identifies with this desire, the sense that if one could just possess the longed-for thing, one would be happy and fulfilled. Ironically, what draws the reader to the story is, in contrast to Al's overriding motivation for acquisition, the desire for a historical identity, the very thing that Al is selling. Led by the narrative to condemn Al for his decision, we are, in this readerly moment, greatly defined by our own desire for a sense of historical connection.

Smith's narrative trick of offering an oral history in written fiction calls attention to the fact that this "history" exists only in the imagined world. Yet, it is the novel's pseudo-historical quality that gives *Oral History* its appeal. Why are we of the contemporary age so anxious to imagine/ understand our history? Perhaps, as the cliché goes, it is so we will not repeat history, or at least the mistakes. But deep down, we know there is more to it than this: We desire to grasp our history because in it lies a sense of our own identity.

Oral history is one process by which to generate that sense of identity. The current interest in folk culture, including folk study centers located on many university campuses, illustrates a sense of urgency in preserving aspects of the past that will be otherwise lost with the passing of generations for whom these practices were part of daily life. Oral histories, including interviews, stories and songs, are recorded or written down to ensure their survival. Yet, once "packaged" and removed from its context, folk culture is no longer that. For example, Joycelyn Hazelwood Donlon illuminates the role of the listener in folk oral-performance. She asserts that "through storytelling, members of a Southern community vigorously reaffirm their connection to each other" (17), implying that the identity generated by storytelling is only possible if the tales are told in the context of a community culturally connected to them. She goes on to suggest that Smith uses narrative strategy to show the limits of "cross-cultural storytelling" (18). Donlon argues that because of cultural difference, Jennifer Bingham's failure to really *hear* the voices speaking to her in the Cantrell cabin "caution[s] members of a storytelling interpretive community to actively receive and confirm these stories, rather than simply to render analysis which seeks to displace the value of the narrative being told" (28). This point is well-taken, in that Jennifer's college project does not render her relatives' stories authentically for her or for any other "outsider" who reads her analysis.

Smith emphasizes with this illustration the great loss implied by the disappearing remnants of folk culture. Significantly, however, she also asserts that for readers to try "to actively receive and confirm these stories" is a self-deceptive exercise. While Smith treats ironically Jennifer's naïve belief that she can visit her relatives and somehow capture her family's history by collecting stories, relics, and photographs, Smith calls us to recognize the distance between teller and listener on more than one level by reminding us that we are even further removed from our folk-selves than Jennifer is. We listen with Jennifer to the ghostly voices of Granny Younger, Rose Hibbits, and Jink Cantrell, and judge her understanding as shallow because until the narrative shifts into her perspective, we have been reading their accounts from a first-person point of view: We have felt privy to an *insider's* perspective on this family's lives. Yet at the end of the story, with the introduction of the tape recorder, asserted retrospectively as the source of the voices all along, Smith reminds us that we cannot have been insiders. For we are reading a tale spun by Lee Smith. Out of our own yearning for a folk past, we have been willing to suspend our disbelief and our analytical impulses almost completely--that is, until we are caught in the act by the author.

Paula Gallant Eckard, echoing Donlon's line of argument, states, "In *Oral History*…the concern is not so much with preserving the past, but with examining, deconstructing, and ultimately redefining the past" ("The Prismatic Past in *Oral History* and *Mama Day*" 121). Here, Gallant emphasizes Smith's insistence that while story is a way to imagine the past, we must maintain our awareness that it is imagined. Further, in addition to Smith's interest in *how* we construct the past, she explores the role and effect of *our desire* for that past. In the process of illuminating the crisis of the Cantrell family, Smith reveals profoundly a crisis common to contemporary society at large, a crisis in which we attempt to locate ourselves as citizens of a global, media- and market-driven world, while grasping at vestiges of our more clearly defined *provincial* identities, which provide a desired, perhaps even necessary, sense of rootedness and belonging.

In particular, through Richard Burlage, Smith's well-educated and cultivated Virginian, we can identify our frustrated need to connect with the past. Early in the novel, defining the goals of his journey from Richmond to the mountain hollers of Appalachia, he explains, "I intend for this journal to be a valid record of what I regard as essentially a pilgrimage, a simple geographical pilgrimage, yes, but also a pilgrimage back through time, a pilgrimage to a simpler era, back—dare I hope it—to the very roots of consciousness and belief" (98). While we tend to find Burlage's self-consciousness a bit tiresome, most readers can sympathize with his plan, albeit a clumsy one, to try and capture something of his roots. Is it not our own desire to do the same that makes the novel so entrancing? What about non-Southern readers, or Southern readers from Mississippi or Louisiana? You may be asking, why are we drawn to lose ourselves in this novel's world of Appalachian past if we are not *from* Appalachia? Well, in some ways, Appalachia is the perfect space in which our historical identities may be imagined. Rodger Cunningham characterizes the region as a relatively unmapped space still open to interpretation. He suggests that in being deemed "other" by America, the South has responded by "otherizing" Appalachia. This dynamic is revealed in the novel's characters themselves. When Jennifer Bingham visits the family homeplace for the first time, to complete her project for a college oral history class, one of her early journal entries reveals just such a perspective: "They still live so close to the land, all of them. Some things may seem modern, like the van, but they're not, not really. They are really very primitive people, resembling nothing so much as some sort of early tribe" (284). Regarding this exoticization of Jennifer's relatives, Buchanan notes,

> *Jennifer's condescension toward her mountain relatives and the fact that this condescension is misplaced are demonstrated when, closing her journal entry with "I shall descend now, to be with them as they go about their evening chores," she returns to the house to find no one doing "chores." Little Luther Wade and Ora Mae are sitting on the porch, Al is puttering in his van, his children are watching Magnum on television. The twentieth century has come to Hoot Owl Holler; the outsider's stereotypes have little to do with reality.* (335)

Suzanne W. Jones comments further on Jennifer's attitude, asserting, "While Jennifer, like some amateur folklorists, tries to prove that a pastoral past still lives in the present, Smith is at pains to paint a more complex picture" (102). Although Jones's comment emphasizes Jennifer's oversimplification of her history, it is important to note the other point implied in Jones's statement: that Jennifer is seeking an imagined pastoral in her "roots."

Like Buchanan and Jones, Cunningham concentrates on the negative implications and effects of Jennifer's stereotyping, examining the impact on the scapegoated culture. Yet, more interesting is an effect he treats as secondary, a *voiding* effect that renders Appalachia as a blank and contemporary society's joy at finding such a blank: "The 'border country' of Appalachia is defined by 'America' in general and the South in particular in terms of a negativity, an area in which dichotomies overlap and therefore cancel each other out—are perceived as blankness" (42). The effect of the overlap is fascinating, if somewhat worrisome. As Cunningham explains,

> *This "mysterious realm," this "terra incognita,"...could only be known by being filled with a version of oneself, and a version deprived of a relation to time... "persistently depicted as stuck in time." The rhetoric survives to this day...in the whole "development" ethos, in whose terms Appalachians cannot have any valid perception of what is in front of them in the present, but instead exist only as a timeless blank to have being conferred on it by whatever intrusion is necessary.* (44)

Cunningham's description of this region as a void emphasizes both its unfixed quality, thus its tantalizing mystery, and its appealing, though fictional, quality of "pastness."

Smith echoes Cunningham's characterization of this region in her reference to the *past* as a mystery: "I guess I see some sort of central mystery at the center of the past, of any past, that you can't, no matter what a good attempt you make at understanding how it was, you never can quite get it" (Qtd. in Buchanan 338). Appalachia, paradoxically defined as *unknowable* and *stuck in time past*, is what makes it such a fascinating and likely region to explore for our own identities.

As a key device, desire is the force that not only attracts us to the novel but also drives the narrative forward. We are plunged into this tale of human desire with the introduction of Almarine Cantrell, living in Hoot Owl Holler around the turn of the century. Drawn first to Red Emmy and then to Pricey Jane, Almarine loves the latter with a rare passion. Granny Younger describes his gaze, turned upon his young wife: "I seen how Almarine looked at Pricey Jane…[I]t was a clear-eyed look, I'll tell you, the way a man don't often look nor love" (64). After Pricey Jane dies, either from "dew pizen" (78) or from Red Emmy's jealously cast curse, Almarine is devastated. Similarly, Pricey Jane's daughter Dory yearns for Richard Burlage until she can stand it no longer and lays herself out on the train tracks, welcoming the relief of death. Sally Cantrell describes her mother Dory's state of longing as something with which the family always subconsciously lived: "Our family was like that, with Mama at the center, not doing anything in particular, but not *having* to either, and all the rest of us falling in place around. Mama was *waiting* somehow, caught up in a waiting dream" (238-9). Sally reflects on her mother's suicide as somehow inevitable: "I think, even then, I knew we had Mama with us on borrowed time, she was so clearly waiting…a place inside her was empty that we couldn't fill" (244). And Dory's daughter Pearl finds herself torn between her socially prescribed duties and her desire for a different kind of life. When Sally asks, "What do you want, Pearl?" Pearl answers, "I don't want anything to be like this. I want things to be *pretty*…I want to be *in love*" (258). Yet, like her Cantrell predecessors, what she pines for is an imagined state. When Pearl does find the romantic love she's yearned for, in her high school art student Donnie Osborne, it ends in tragedy, Pearl dying of complications at the birth of Donnie's and her child and the baby dying soon afterward (275). Sally says of her sister, "What Pearl wanted—I've always thought this—was several lives. One was just plain not enough" (242).

Richard Burlage's journal narrative, placed among the tape-recorded voices, is driven by desire, as well—not for Dory, but for experience. He looks backward to find a more "real" experience than that of his present world in Richmond. To Richard, this Richmond world seems lifeless and artificial, and as a product of this world, he believes he is prevented from truly experiencing events, people, and places. But his plan for overcoming his academic approach to life ultimately fails. In fact, it has the opposite effect of what he intended. During his first stay in the mountains, he creates a simulacrum of the folk culture as he renders it in his journals, and later in his photographs. On his first trip, he experiences mountain life as an observer. Even when he attempts to engage in a salvation experience at a Freewill Followers Church meeting, he cannot shake the feeling that he is not directly experiencing the conversion: "[E]ven at this crucial moment, I remained a sojourner still—it pains me deeply to admit this—I was observing my actions even as I performed them" (154). Years later, he tries again to open himself up to the place. Yet, he states, upon his return to the hollers, this time with a camera, "My vantage point on the hairpin turn of Hurricane Mountain, facing this coal camp, made me feel omniscient: I could view it all and view it whole, the people tiny, not real people, not at all, the cars and trucks nothing but toys" (225). He traps himself and others with his simulacra. Not only does he find it impossible to escape his artistic versions of experience, but he also imposes his own meaning on those he writes about and photographs, publishing his work as evidence of the region's quaint cultural character.

As disdainful as we may feel regarding Richard's falseness, we, as readers, become guilty of the same tendency toward construct. As the objectified rather than the objectifier, Red Emmy gets trapped into an identity created for her by those who, unlike her, are given a voice in the narrative. Granny Younger, the respected midwife and healer of the community, calls her a "Child of the devil" (52). The identity given to Emmy by others supersedes her probable complexity as an individual. She becomes, for them, the witch Red Emmy. Lynda Byrd emphasizes that the community's psychological fears drive the story that becomes Red Emmy's:

> *Even though Almarine appears happy, always smiling as he does his chores, Granny insists that he is "bewitched" and no one in the holler will have anything to do with him.... When Almarine becomes ill, Granny has a ready explanation: "I knowed what was happening, of course. A witch will ride a man in the night while he sleeps, she'll ride him to death if she can."*
> ("The Emergence of the Sacred Sexual Mother in Lee Smith's *Oral History*" 124)

I Have Been So Many People

Cunningham asserts that simulacra like those created by Richard Burlage and Jennifer Bingham are subverted to some extent by the form of the novel: "*Oral History* is indeed, as its title implies, a pastiche of different voices—so different that we eventually suspect that none is reliable. And as soon as we do so, we begin to hear the whispers of the authentic voice underneath, a voice made present by erasure—a blankness made articulate" (48). Here Cunningham seems to assume that readers are above succumbing to the seductions of the narrative itself. Perhaps he gives us credit for admitting to ourselves, as we read, that Dory and Emmy (and we) are unknowable by the process in which we are engaged. I am, however, a bit skeptical of this optimism. If we are drawn to the story by our deep-seated desire to belong historically, we may be quite willing to see Emmy as Granny Younger represents her, as the witch who works the farm by day and rides Almarine slowly to death by night. For it is the folk history we are after.

But Smith forces us back into confrontation with ourselves and our desire in the end of the novel when it becomes clear that the voices we have been "listening to" are the ghost-voices captured on the tape-recorder of Jennifer Bingham—great-granddaughter of Pricey Jane and Almarine Cantrell—who is searching for remnants of her ancestral heritage. This narrative's strategy of the tape-recorded voices, which catches readers by surprise, seems so ridiculous that at first, they might even be angry. I, for one, felt cheated when it dawned on me what Smith was up to. I reasoned that I could have suspended my disbelief easily enough to drift through time at the whims of the various narrators, several of whom are already dead at the time of the frame narrative. But it seemed Smith asked too much in expecting me to buy the trick of the tape-recorded ghost voices. Despite H.H. Campbell's comparison of *Oral History*'s "spookiness" with that of Bronte's *Wuthering Heights* (143), the "hants" of the Cantrell cabin simply do not generate the quality of goose bumps that are produced by the ghosts of *Wuthering Heights*. With much close-reading, however, I have come to see a trick *behind* Smith's ghost trick. The postmodern twist of her ending has the effect that most metanarratives do: They force us to look in the mirror, to reevaluate our expectations of the narrative and our reasons for reading.

The novel, Smith reminds, is no less simulacrum than Burlage's journals or Ghostland. The past has literally been lost, but the mainstream culture is wild to recreate it. Jennifer, who has lost her ancestral past and the sense of identity it provides, yearns to recapture it, just as the public does. While this novel is a wonderful and rich simulacrum, much more satisfying than Ghostland, it is, Smith reminds us, a representation just the same. Perhaps Rosalind B. Reilly says it best in "*Oral History*: The Enchanted Circle of Narrative and Dream":

> *Granny's metaphor [of the landscape] suggests that the storyteller, like the healer, ministers to "sufferers" of a sort, those who are so hemmed in by their experiences [or their ideologies, I would add] that they cannot wander freely throughout the landscape of the imagination in order to find what they "want the most."* (80)

Smith helps us recognize, if we listen closely, that we are yearning for a folk past and prods us to consider *why*.

Is Smith condemning our yearning for the past, our constant engagement of creating/imagining the past? Absolutely not. For one thing, stories are as important to Smith as folk history and respect for regional integrity. Second, she is *intrigued* by the desire that causes us to seek out our roots and to construct identities that reconcile us to our perceived pasts, specifically our folk pasts, which are often more rich and interesting than the commodified histories publicly written, histories lacking in the details that tie us to region and to specific family and community cultures. But Smith insists that as we forge these connections, imagining coherent and meaningful identities in the process, we must remember that we are creating worlds and imposing meaning, else we may trap ourselves and others into those reductive and sometimes oppressive visions, as Richard Burlage does.

It is no coincidence that Ora Mae fixes on the golden earrings as symbols of the family curse. They are literally objects of desire, and further, they are fixed and unified in their circularity. They represent both the desire that makes us human and the dangers of becoming trapped in that desire.

Chapter Four
The Culminating Self in *Family Linen*

Then Sybill remembers something else, something perfectly dreadful: soon after that night, she took the long black flashlight from the high hook in the toolshed and shone it down the well. She remembers seeing her father's face there just for a minute, beneath the shiny black water. She saw his high pale forehead, his open eye. (Family Linen 34)

In *Oral History*, Smith portrays our thwarted desire for access to the essence of reality, including the self. In exploring that desire, *Oral History* acknowledges our tendency to see the past as an antecedent to the self and to seek/construct the past in order to better understand ourselves. Yet as Jennifer Bingham's rendering of her family's history shows, the connection between past and self is not a simple, linear, cause-and-effect relationship. The Cantrell history is portrayed as an elusive, and often moving, target.

In her sixth novel, *Family Linen* (1985), Smith further investigates this issue, this time considering the dangers of repressing, or even denying, one's history, elusive though it may be. In this novel, Smith asserts that although a true and complete history is impossible to ascertain, refusing to acknowledge our past and its impact can be downright crippling. For Smith, memory is not a simple reflection of history; on the contrary, it is a fluid and often unreliable mediator between the past and present. Yet, the role that memory plays in self-construction is crucial to the evolution and the healthy function of the self. In *Family Linen*, most members of the Bird/Hess family struggle between their desire for a past they can be proud of and the subconscious need to know the truth, and through their struggles, Smith reveals the pitfalls in negotiating the space in between. Further, with Candy, the Hess child considered as unspoken "outsider," Smith introduces the notion that *artistic endeavor* can, even in this ambiguous space, provide a way to self-acceptance and healthy community.

Family Linen portrays the Bird/Hess family of Booker Creek, Virginia, struggling to sustain both their place in the society of this small Southern town as well as their individual hopes that they will each be "okay." Except for Candy, they are alienated from one another and from themselves by their repression of the difficult and "unsavory" elements of their past. Though repression as a coping strategy seems to have worked well enough until now, the oldest Hess daughter, Sybill, has begun to suffer the effects of denying the family history. Seeking an understanding of and relief from her recently developed headaches, Sybill watches a television program on Alzheimer's disease one day and notes the coincidence of its symptoms with her own: "Alzheimer's disease causes memory loss, confusion, speech impairment, and personality change," she considers (15). Typical of her usual dismissal of anything unpleasant, she rejects the possibility that it could be causing her own symptoms, telling herself she "doesn't believe in Alzheimer's disease" (15). Yet, when she can no longer tolerate her intensifying headaches and debilitating anxiety, she finally seeks medical help. Understanding that her problem is likely stress-induced, her doctor sends her to a hypnotist. Sybill is hesitant to follow his advice to undergo hypnosis, since she "regarded her unconscious like she regarded her reproductive system, as a messy, murky darkness full of unexplained fluids and longings which she preferred not to know too much about" (4), but the headaches do not abate, so she eventually acquiesces. At this point, Sybill recognizes her inability to continue repressing what lurks beneath her carefully constructed self-narrative; she cannot tolerate the pain being generated by these buried memories.

Though Sybill warns the hypnosis therapist Bob Diamond that she's "not here for any psychiatric" (11), he argues, "I feel that there are certain facts which might be brought to light which could, perhaps, help us to understand the origins of your pain" (12). Despite her psyche's longstanding resistance to remembering the images of a particularly momentous night in her childhood, her hypnotic state enables Sybill to see clearly "the figure out in the yard, with the long-handled ax upraised, and [to see]… her bring it down again and again into the man who lay on the hillside in the streaming rain, the washing mud" (31). When Sybill comes out of the hypnosis and begins to consider this vision as the repressed memory of her mother's murder of her father, Jewell Rife, the second memory comes to her, the image of her dead father submerged in the well.

The strength of Sybill's newfound memories drives her to overcome her usually prim manners and to blurt her story to the rest of the family. While the habitually shrill Sybill does not hold great sway over her brother and sisters, her announcement, along with her mother's stroke which has occurred almost simultaneously with the hypnotic episode, launches the Bird/Hess family into a confrontation with their past and into an ensuing phase of self-reflection.

Although the family has always had a certain standing in Booker Creek, this reputation has reflected only the surface of their lives. In the course of the novel, through shifting narrative perspectives, we see the family at various moments in their history. The narrative begins in the late twentieth century when Sybill's mother, Elizabeth Hess, is old and unwell; her first husband, Jewell Rife, and her second husband, Verner Hess, are both dead, and Elizabeth's grown children are gathering to do what they can for her as a result of her stroke. Through the characters' memories, recollections of family lore, and Elizabeth's long-buried journal, we learn of the family's beginnings, as well. Elizabeth's journal tells of her parents' (the Birds') arrival in Booker Creek and the challenges for the young Elizabeth when her mother died; through Elizabeth's sister Nettie, we glimpse the intense, but short, marriage of Elizabeth and Jewell Rife, and we learn of the time following Jewell's disappearance, when the family struggled to survive, as well as of Elizabeth's eventual marriage to Verner Hess. As we piece together these memories and stories with narrative portrayals of recent events, we begin to understand that the Hess family's current relations to each other are complicated not only by Jewell's sexual relationship in those early years with Elizabeth's mentally challenged sister, Fay—producing the child Candy—but also by Jewell's subsequent disappearance when his and Elizabeth's two children, Sybill and Arthur, were still very young. Although Elizabeth's second marriage to Verner restored economic and social stability to the family, soon increased by the birth of two more children, Myrtle and Lacy, Elizabeth's determination to preserve her family's reputation over the years has come at a price: the repression of unsavory details of their history has left them in fragmented relationships and hampered their psychological growth. The history-based identity she has passed on to each of them is inauthentic and fragile, leaving them unable to negotiate the challenges they face in their complicated adult lives.

Yet, as the family members negotiate their past, struggling to maintain stability, it is clear that for each of them, Elizabeth herself, both real and imagined, is an important orientation point. Each of the daughters feels that she was the closest to Elizabeth, that she understood her mother best. Yet each also feels that she never quite met Elizabeth's standards, and this feeling has caused much anxiety for Sybill, Myrtle, and Lacy. The only son, Arthur is perhaps even more extremely affected by his relationship with his mother. Told by Elizabeth the last time he saw her that "he needed a haircut and that he had been for her a source of constant pain"

(104), Arthur has lived an aimless and dissolute life, developing a kind of numbness regarding her, even upon her death: "Mother is dead, he tried to think, but so far this meant nothing" (107). Harriet C. Buchanan notes, "By Lacy's section of *Family Linen*, the reader is well and truly caught up in the kaleidoscopically shifting patterns of the different family members' perspectives on the life, illness, and central mystery that is Miss Elizabeth" (340). Seeking a self that enables movement forward, each sibling still yearns in retrospect for Elizabeth's love and approval. Facing the heretofore curtained past is one step they must take in pursuit of healing.

Since Elizabeth never recovers from her coma, Sybill never receives confirmation that her hypnosis-induced visions were accurate. The unanswered question regarding her father's murder haunts Sybill and necessitates examination of her self, her plans, and her relationships.

Similarly, the relationships of the other characters to their pasts are selective and sticky, even for Lacy, the youngest, who never even knew Jewell Rife. Upon coming home to the house on the hill to see her sick mother--never considering that she might have to attend Elizabeth's funeral during this trip--Lacy feels conflicted; she recognizes that for her, "The house was symbolic of so many things" (66). She has stayed away for years, after all. Although she has a reputation for being a "free spirit," as of late she has become somewhat directionless. She never finished her dissertation, even after completing all the coursework for her doctorate. Back in Chapel Hill, she has also been trying to deal with her husband's recently leaving her for another woman, so delving into her family's past at this point in her life is not easy for Lacy. Considering why she left Booker Creek in the first place, and why she has rarely come back, she surmises that her siblings and she needed to gain some distance from their upbringing and the homeplace itself in order to establish independence: "Lacy thought of how they had all preferred, finally, in different ways, terra firma. How they had all chosen to come down the hill" (66). The family home is symbolic for Lacy of the family's history and of their reputation as a "good" family in Booker Creek, as well as of the burdens of this role. As the youngest, Lacy understands less than Sybill does about why her family is not as close as they might be, yet she recognizes the silences and tensions. Elizabeth's death, however, has brought them together to re-explore the house in which they grew up and their shared past.

In the process of this physical and psychological exploration, Lacy finds the journal kept by Elizabeth as a young woman. In this diary, Elizabeth describes her own past as a metaphorical house that provokes emotions resembling those of Sybill and her siblings about their own homeplace: "I approach the Past as a young maiden, bearing a candle, might approach a deserted mansion deep within the Enchanted woods... This mansion is no place for the faint of heart, no place for the unprepared" (168-9). The tone of Elizabeth's musings indicates that while she wants to preserve and understand the story of her early life, she dreads the pain inherent in the exercise.

From this journal, Lacy learns that in the early 1900s, Elizabeth's father, Lem Bird, built the house originally for her mother, Mary Davenport, upon this hill because "only the prettiest spot would do" (174). Furnished with "[d]raperies and furniture from Richmond" (175), the house was especially designed for a "lady" and paid for from the profits of Lem's lumber business. The vision Lem meant to realize in this endeavor would become the family's vision of itself, regardless of the contradicting developments. Elizabeth internalized the notion and would foster it in her own children, who would be both driven and burdened by it. Later, when Mary died and Lem's business failed, Elizabeth decided that she would need to become the "Lady of the House" (190), though it is clear that in taking up these duties as such a young woman, she could not have anticipated that to maintain the apparent nobility of this household, she would have to endure an unfaithful and abusive husband in Jewell Rife as well as the loss of the family fortune, cope with Jewell's early death, and then raise five children with Verner Hess, with carefully guarded grace and stoicism. As a result of this commitment, she has preserved the family's position and home but set in motion a repressive and isolating model of self-definition in the process.

Lacy recognizes that there is much history buried in this place, literally. As she tries to remember what it felt like to live here as a child, she remembers the sense of mystery it held for her. She asks herself, "Can you find the secret here, at the heart of the house?" (165). In finding her mother's journal buried in a drawer, Lacy feels she has found a treasure: the story written there fills in some of the gaps of her mental portrait of her mother. Yet, as Elizabeth's carefully rendered version of herself and the family's story, the journal fails to answer some of the most pressing questions about this family's history and thus to help Lacy understand herself and what future she should pursue.

Unlike her prodigal younger sister Lacy, Myrtle intends to carry on what her grandparents began. Identifying strongly with her image of their stalwart mother, Myrtle would rather not dwell upon the ways her family has veered from propriety. She does not want to consider the accusations Sybill has made, that their mother may have murdered Jewell Rife. She finds it difficult enough to deal with her own recent unconventional behavior. Myrtle, married to Don Dotson, dermatologist, has been having a secret affair with Gary Vance, exterminator. Through this affair, Myrtle has experienced situations she avoided for many years as a "proper" daughter, wife and mother: "[O]ne of the things Myrtle will always remember... [is] how hot it is in summer, how cold in winter, at Gary's house. Her own house has central heating and air, so she isn't used to these drastic changes. She isn't used to perspiring or shivering, since she never does either one of those things at home" (44). The comfort of her "genteel" position has kept Myrtle from pain, to a great extent, but it has done so by insulating her from and numbing her to many of life's complexities. Myrtle's repression of her primal urges, as a result of her commitment to maintaining her family's reputation, has heightened her subconscious desire for a more liberated situation. "[S]he has days when she feels like her whole life is a function of other people's" (38); acting constantly as a respectable wife and mother has alienated Myrtle from a more authentic sense of herself. When Elizabeth has her stroke, Myrtle is disturbed to feel almost glad, somewhat freed from the sense that "No matter how good we were, it was never good enough" (49).

Stimulated by a similar desire to *feel* his life, Don himself engages in what has become a long-term, though sporadic, affair with Myrtle's half-sister, Candy. When Don is with Candy, he feels he can temporarily let go of his filial obligations and relax. Even in light of these mildly subversive actions, though, Myrtle and Don are devoted to each other as well as to their sometimes difficult children. Further, they have no intentions of permanently abandoning their carefully crafted public selves. It seems fitting that Myrtle and Don have inherited Elizabeth's house upon her death, though Myrtle is unsure what may be unearthed as they update the property. As the bulldozer digs out for a pool and, Myrtle hopes, proves for all of them that Jewell Rife is *not* buried where the old well used to be, Myrtle "will put the bulldozer right out of her head and not mention it once. Nobody will mention it, and then they can gently say to Sybill, 'Okay, look, honey. There's nothing there'" (159-60). Myrtle refuses to allow her affair to change her sense of who she is, a self she deems necessary to her narrative of the family's history. We are left to wonder if the memory of her affair with Gary will provide a sufficient balance for Myrtle against the constant job of maintaining the family's social position, a job that requires denial of self and of the full details of the past.

Arthur, the only boy and, in Sybill's opinion, the black sheep of the family (26), appears less burdened by the family's expectations than his sisters, but his life has definitely been shaped by his sense of not belonging, by his own troubled relationship to the past: "Arthur doesn't give a shit about much of anything now" (81). He drinks too much, has gotten into trouble with women, and currently house-sits for his living. He earned his mother's disappointment early on, though Verner Hess tried to help him. When Arthur had trouble holding down a job, Verner even invited him to work at the dime store. "'I don't know,' Arthur said. Verner Hess was not his own daddy, and he was thinking about that. Arthur wished he was. But he was not, and somehow that had made a difference in his life" (97). Now in his early 40s, Arthur does not know what happened to his father, Jewell Rife. "Arthur wondered where his own father lay, if he was buried, or where he lived. Arthur wondered if his father ever thought of him" (105). The lack of closure regarding Jewell's disappearance has left Arthur unable to construct a stable self or meaningful, enduring relationships.

Even their aunt, Elizabeth's unconventional sister Nettie, perhaps the least pretentious member of the family, is disturbed at the sight of the bulldozers digging into the old property for the new pool: "All that earth looks startlingly red and raw, against the green" (220). Yet, it is through Nettie's perspective that we discover many of the details that have been expurgated from the family's history. She knows, for example, that Candy is the child Jewell sired with Fay, a child whom Elizabeth insisted on raising as her own. Nettie has chosen not to air out these details publicly. Her estrangement from Elizabeth when they were young women has caused her enough pain that she has decided there is no point in "hanging dirty laundry on the line" (260). Yet, we know, from Elizabeth's journal, that Nettie suffered too, as a result of Elizabeth's insistence on preserving the family's name. Elizabeth writes that soon after their father's death, Nettie "conceived a wild scheme" (201) to sell the house. Though they badly needed the money, Elizabeth would not consent and writes,

> *I could not in good conscience condone such speculation. Nor could I stand to see us sink so low. I remembered our Mother, and the manner in which we had lived during our Childhood, and my heart was grieved past bearing it. She would not have wished us, ever, to leave this House, this situation.* (201)

Elizabeth's perception of her family's nobility of character remained intact, necessitating a split with Nettie, whose honesty threatened it. As a result, Nettie feels that the cost of honesty is sometimes not worth the gain.

The bulldozer does uncover human bones, however, and news comes almost simultaneously that Fay has been found dead. Nettie deduces, from the initial information reported to them, that Fay climbed inside a hot "junk" car that lacked sufficient oxygen, intending finally to take the trip to Florida that Jewell promised her so long ago. Further, based on this quickly-formed conclusion and on the undeniable fact of the bones just exhumed, Nettie decides, and says aloud to the family members standing near her in the yard, that it must have been the disappointed Fay who killed Jewell all those years ago. Now deceased, Fay can neither confirm nor deny Nettie's theory, and the family's silence in response to Nettie's announcement implies that even if Fay did kill Jewell, it was not her fault after all, not in her childlike state of mind. Whether or not Fay murdered Jewell, this version of the story is easier for the Hesses to reconcile themselves to than the alternative. The possibility that Elizabeth killed her husband has already destabilized their identities, even if only temporarily. In their struggle to regain balance, they accept Nettie's theory. These siblings have experienced the pain and disorientation brought on by confrontation with the past, and though they will not resolve all of the discrepancies in their memories and their collective history, this process has fostered them toward healthier relationships and, at least temporarily, healthier selves.

Perhaps fittingly, the novel begins with Elizabeth's coma and death and ends with a wedding and an impending birth. In the middle of the confusion near the pool site, Myrtle's daughter Karen arrives home from college, pregnant, with her boyfriend in tow. Karen's unconventional choices have been a challenge for Myrtle up to now, but this sudden appearance, juxtaposed against the scene of the bones so recently dug up, seems to remind them all, including Myrtle, that they're all alive. Having spent some necessary time facing their difficult past, the family welcomes a chance to celebrate life. As they make ready for a wedding, Lacy feels renewed somehow. Her husband Jack has come to Booker Creek and brought her flowers, and she relishes the company of her children, who are here now for the big event. Further, she's feeling optimistic about her future: She

> *can even see the butterflies on all the blue flowers by the fence, and she can see all of them, herself included, this odd gaggle of disparate family teetering here on the brink of the past while all around them, it's just another pretty day. Full June. Suddenly, for no reason at all, Lacy feels like writing her dissertation.* (271)

Significantly, Sybill's headaches have ceased. And though she has for so long thwarted her own chances of becoming intimate with other people, she is suddenly giving herself permission to engage wholeheartedly in the wedding and even to hug her brother Arthur. When she plugs in the pool lights which "[give] off a soft blue mysterious glow right in the middle of the whole reception" (287), Sybill is delighted at everyone's awed reaction. She "goes to stand at the edge of the patio, hands on hips, and cries a little, there where nobody can see her. This is not sad crying, though; and it comes up from the bottom of Sybill's heart, for just no reason at all" (287). Though her visions have not been confirmed as accurate memories, she has allowed her subconscious some necessary oxygen, and seems to be the better for it.

Even Arthur is able to put his insecurities to rest as a result of confronting the "skeletons in the closet." He seems "ebullient" (287) now; he has a new lease on life, with plans to turn his Aunt Nettie's old One Stop shop into a hotel and run it with his new love, Mrs. Palucci, who was his mother's nurse in her last days. He still feels fleeting moments of hatred for Sybill, who "has seen his father's face, even under water," but he appears to be resuming his own role as a father, with his own daughters and with Mrs. Palucci's son, Buddy, a sign that his disorientation, his need for self-numbing, is abating. He recognizes his own newborn optimism: "There ought to be some interesting moments, over the years" (289).

Candy, who has remained something of a mystery until late in the novel, is illuminated in this last stage of the story. The narrative reveals that she has always been more self-accepting than the other Hess children. While she still does not know she is the child of Fay and Jewell, she has never quite given in to Elizabeth's expectations: "The trouble with Candy is, she's always done exactly what she feels like, that's just the way she is" (111). Perhaps as a result of her being this way, Candy has not resented Elizabeth like her siblings have, although she was often irritated by her. Candy eloped young and had two children, one of which was with her now ex-husband husband, Lonnie Snipes, the other one born out of wedlock. She knows she "turned out different from what her mother hoped" (122). Yet, unlike her siblings, Candy sees Elizabeth beyond her role as mother and judge. A hairdresser by trade, Candy agrees generously to prepare Elizabeth for the funeral. As she works carefully on her mother's body, she thinks,

> *Miss Elizabeth was a lady, and she looked like a lady in death. She had lived, Candy reckoned. She had had children, she had felt things, thought things, she had died. She had loved one man, and another man had loved her. It's hard to say which of those conditions is better or worse. Sometimes neither one of them happens. That's probably the worst.* (121)

Having refused the "mantle" of the family position from the beginning, Candy has matured without its crippling effects and has developed a more flexible sense of self than her siblings. This alternative identity has enabled her to relate to others with less pretense and thus with more authenticity. Candy has had her share of difficulties, raising her two children alone. But, she connects with her mother in her understanding that, unlike Sybill in her long-standing resistance to emotional risk, they both embraced life, pain and all.

There is an additional factor, though, that enables Candy not only to embrace her past but also to sustain a flexible self, capable of withstanding life's narrative-changing obstacles: her art. While Candy and Elizabeth both found expression of their artistic impulses, Elizabeth through her journal and poems and Candy through her hairstyling, Candy's work goes an important step further than Elizabeth's, drawing together the lives around her, including that of the deceased, Elizabeth. Debbie Wesley explains:

> *Candy's apartment above her shop in the center of town suggests her ties to her community. From her window she can see and participate in the daily happenings within the town. Candy is involved in the most important rituals of the community. She might style the bride's hair for her wedding day or give a new mother a hair-cut to suit her new role in life. And it is Candy, the black sheep daughter, who carefully and lovingly does her mother's hair and make-up after Miss Elizabeth's death.* (93)

Candy's art is manifest in relations to the people in her life and in her sense of self, as well. Katherine Kearns notes that Smith characters like Candy have achieved an artistry that is significantly consumable: "What these artists/muses make is at once both highly tangible and ineffable. Having struggled toward wholeness, they can provide for others, and with no defensive agenda, they are open to epiphany through the commonplace" (185). Candy's art offers her a connection to her mother, both because of their shared artistic impulses and because Candy's hairstyling is inherently communal and intimate.

The Hesses' intense encounter with each other, the past, and themselves proves difficult but important for each of the characters. By the end of the novel, they are perhaps no closer to a clear vision of who they are, or who their mother was, but they have grown somehow by facing their fears linked to the past. The necessity of this self-examination seems to be Smith's point. Harriette C. Buchanan asserts, "The exploration and representation of those mysteries, of those human lives, are the artist's endeavors, and she [Smith] will continue to present them to us as her characters, in their own ways, tell us their stories" (342). Thanks to the past's "unearthing," the characters seem, at least temporarily, liberated. Candy gives Sybill a new hairstyle, a bold move for Sybill, and as a result, Sybill "looks ten years younger" (277); in addition, Candy does Myrtle's color with more ash blond because Myrtle has decided to let the gray start to show, an equally courageous decision for Myrtle (278). On the day that Karen marries her lover Karl, it is not likely that the novel's characters have put their past to rest; however, Nettie's grown stepson Clinus has posted a message on the

One Stop marquee that applies perhaps to them all: "Karen and Karl, today is the first day of the rest of your life" (290-291). If the Hesses are plagued by their indefinable relationships to each other and to the past, they have at least brought those factors to the conscious level for an airing out, and they seem the more resilient and happy for it, more equipped to face the future.

Chapter Five
The Protean Ivy in Fair and Tender Ladies [1]

Having introduced an element of metafiction into her body of novels with *Oral History*, Lee Smith has employed this postmodern strategy in several works since, to emphasize humanity's common practice of creating ourselves by way of narrative. Ivy Rowe, the captivating protagonist of Smith's epistolary novel *Fair and Tender Ladies* (1988), writes her letters for a number of reasons. She uses them to communicate important news, to "keep in touch" (82) with family and friends, and to record events and feelings. Beyond these functions, however, Ivy's letters serve as a mirror in which she can see herself, through which she can better understand herself. This self is not a permanent or static self, revealed to her bit by bit in her letters. Rather, it is a fluid self, comprising many identities; as Ivy writes late in life to her daughter Joli, "I have been so many people" (266). Although as a Southern, Appalachian woman, Ivy experiences pressure from a number of dominant ideological influences, she resists absolute definition by any one of these ideologies. This is not to say that Ivy is immune to the pressures they exert on her; certainly, her expressions often echo the terms of those ideologies. However, in the end, Ivy's identity is too protean to be fixed by any particular system. This fluidity results in large part from Ivy's habit of deconstructing, through her letters, systems which threaten to entrap her.

To describe Ivy's deconstruction of these systems is not a simple task. She does not consciously and systematically tear them apart; she does not even subvert them to the point that they ultimately have no effect upon her. Instead, Ivy, expressed to us through her letters, fluctuates between

---

[1] Much of the content of this chapter has been taken from my article, "The Protean Ivy in Lee Smith's *Fair and Tender Ladies*" originally published in *The Southern Literary Journal* 30.2 (Spring 1998): 76-95.

acceptance and rejection of the ideologies that influence her. If we can view acceptance and rejection as opposite ends of a spectrum, however, we can see that although Ivy moves back and forth on the spectrum rather than progressing in one direction from acceptance to rejection, she ultimately deconstructs all the ideologies that work upon her. For example, while she still loves romances at the end of the novel, she exhibits an awareness that the formula behind them is an artificial one.

At the same time, Ivy does not *set out*, in writing her letters, to deconstruct the ideologies that dominate her culture or herself. She writes instead for various other reasons. Anne Lieberman Bower notes that Ivy uses the letter both to establish presence, as when she writes to "hold together a dispersing family and to continue traditions that are fading" (33) and to emphasize absence—which desire is exhibited in letters like the one to her pen pal Hanneke, in which she expresses her anger that Hanneke has not written back, and those to her sister Silvaney in which Silvaney's absence is necessary to Ivy's self-exploration (46-7). Ivy often writes to convey information. Also, as her letters to Hanneke reveal, she is a lonely girl and yearns for a way to break the isolation of her existence up on Sugar Fork. Most significantly, however, she admits that her understanding of herself is limited and that her letters give her a way to understand her experience, writ not only to remember her experiences and feelings but also to examine and shape them. She explains this need in a letter to her friend Violet Gayheart: "Sometimes I despair of ever understanding anything right when it happens to me, it seems like I have to tell it in a letter to see what it was, even though I was right there all along!" (181). Later, in a letter to Silvaney, who has died by this time, Ivy says, "You know I have always got to write my letters, and think about what's happened and what I've done" (245).

Addressing Ivy's need to know herself, Dorothy Combs Hill suggests that

> *[t]he wound that the whole important body of Lee Smith's work redresses is the terrible cultural wound inflicted on creative women that keeps them from understanding themselves and even denies them any access to themselves. They have no access to themselves, neither in their own unfree imaginations nor in collective institutions. There is no form in which they can recognize themselves, no form in which others can recognize them, certainly not without unsexing them or deforming them.* (xvii)

Herion-Sarafidis addresses this problem for Smith's characters, specifically for Crystal from *Black Mountain Breakdown*. Herion-Sarafidis explains,

"In looking for herself, Crystal is never real to herself, she is only real as she is mirrored, as crystal, in how others see her and treat her" (15). Hill and Herion-Sarafidis acknowledge, however, what I will argue in this chapter, that Ivy, unlike her predecessors in Smith's works, finds a way to see and to recognize a more substantial self. While Hill considers Ivy's retrieval of rejected ancient archetypes as the key to finding herself, I would argue that it is Ivy's creative use of the letter as mirror that allows her this construction of a liberated self.

Ivy's experience with actual mirrors repeatedly reveals the seeming lack of any way for her literally to see herself. For instance, when her uncle Revel takes her and her sister Beulah to town to get new dresses, Ivy tries on her dress and then attempts to get a look at herself in the store mirror: "I was looking in the mirrer and trying to see, but it is dark in the back of that store and the mirrer is wavy" (70).[2] Later when Ivy and Beulah catch a glimpse of themselves in a store window, Ivy does not even recognize herself at first: "Now who is that? I thought to myself as we turned the corner and waited in front of the pharmacy to cross the street. And Molly, it was us! Us in the winder looking like movie stars, me too" (70). The next mention of a mirror is not until much later in the novel, and still the object is a novelty to Ivy. She writes to Silvaney about her experience with her lover Franklin Ransom: "[H]e turned on the bedside light and got me to look in the mirror door. I had never seen a mirror door before. I had never looked at my whole body all at one time." It is an affirming experience for her since it results in her discovery that "I am beautiful!" (163). In her letter-writing, she is attempting something similar to looking in the mirror. Explaining that she needs to write things in order to understand them, Ivy reveals her need to be able to step back from an event, a feeling, or an image in order really to be able to look at it. The letters function for her as a way to do this, and in the process, they foster in her the possibility of fluidity, of her identity as a fluctuating self, rather than a fixed one.

A manifestation of her fluidity, Ivy's constant authorial movement makes it impossible to define her position on virtually any issue. That is, in one letter, she may express a perspective that in another letter she contradicts. Since the novel is a collection of letters (each a fragment of the whole collection), Ivy can depart from the notion of the unified narrative, producing at the same time a whole work comprising various impressions, the combination of which cannot ultimately be contained by any particular

---

2 The misspellings are Ivy's, reflecting her lack of formal education as well as her dialect. Further, certain words, phrases, and sentences are underlined in some of this chapter's quotes from Ivy's letters. This strategy is also Ivy's and seems to be used for emphasis.

ideology. Composing complex and contradictory impressions is the most obvious way Ivy uses the letter to deconstruct the ideologies that impose upon her. For example, she wavers between regard and disgust for the conventional Christian God, rendered to her primarily by Southern evangelical preachers. In her sixth letter to Hanneke, angry at her family's poverty and her father's illness, and at Hanneke's failure to write back, Ivy unleashes her ire: "I know I am evil and I wish evil for you too. Mister Brown told us one time that God is good, but He is not good or bad ether one, I think it is that He does not care" (17). Her statement that "He is not good or bad ether one" implies that Ivy does believe in a male God, a God who, although He may not care about what happens to His creation, seems to have the ability to interfere with it at will. Indeed, she criticizes His divine actions: "no I do not pray, nor do I think much of God. It is not rigt what he sends on people. He sends too much to bare" (32).

From the beginning, then, Ivy doubts the innate goodness of the Christian God, thus undermining to some extent this Christian ideology's power over her. Yet, there are also times when Ivy's letters indicate that she experiences some fear of God. She writes Beulah, for instance, about how she almost got "saved" at "the meeting" where Sam Russell Sage has been preaching. Ivy feels that if it had not been for her teacher Miss Torrington's almost fainting, she would have responded to the invitational, and afterward, as she writes her letter, she worries: "I have not been saved yet, so I hope I will not die anytime soon!" (94). Afterward, however, she discovers Sage's hypocrisy and returns to her initial position on the nature of God, claiming that "if Sam Russell Sage is who God has sent, then I don't know if I even want to be saved ether" (97).

Later in the novel, after Ivy's brother Garnie has grown up and become an evangelist much after the model of his mentor Sage, Garnie becomes determined to bring Ivy under the umbrella of his own hypocritical program. Ivy begins this story in a letter to Silvaney with "I have to write to you, for I can tell no one. This is the story of how I was not saved" (254). Ivy then writes of how she attended one of Garnie's services but that her mind wandered, and she spent her time concentrating on who was at the meeting rather than on Garnie's message. She left the service without having been converted, though she does consider in her message to Silvaney whether it was wise to do so. On the other hand, after her husband Oakley expresses his desire for her to "believe," she considers the possibility that "[m]ay be it is finally time" (258). She has blamed herself, in particular her affair with the bee man, Honey Breeding, for her daughter LuIda's death, thus assuming some sort of cosmic justice system. She deliberates on the idea that she is a sinner and that God can "work in mysterious ways" (259), yet she simply cannot buy fully into the ideology that gives rise to so much hypocrisy. She reflects on Garnie's subsequent visit to the house at Sugar Fork to urge her once again toward God:

> *So it hit me, there on the hillside, <u>This could be it, after all these years. It could be God speaking out through your fat little brother Garnie, and why not? Stranger things have happened.</u> But because I am so contrary, Silvaney, another part of me said, <u>Well, if this is the vessel God has picked to carry his message, then it is a mighty damn poor one!</u>"* (259)

Ivy's wavering confidence in this system of salvation, particularly as it is represented by Garnie, is finally shattered when, in the name of "righteousness," Garnie condemns her pride and begins to remove his belt, "with a furious face and drooling spit and panting out loud like a dog" (262). While Ivy assumes his plan is to whip her with his belt, his intentions are left to our imaginations since Oakley arrives in time to beat Garnie until he can only stumble back to his car.

Exemplifying Ivy's constant shifting, the contradictory perspectives she exhibits in regard to Southern evangelical Christianity are juxtaposed against a different type of transcendent experience, also described in her letters. Ivy experiences a sort of spiritual life in connection with nature and, significantly, shaped by Ivy herself. In the letters describing this alternative spirituality, it is as if Ivy believes in the Christian God but makes a conscious choice to exist outside of his kingdom or protection. Instead, she unexpectedly reaps spiritual fulfillment during moments of her daily life. Interestingly, several of those experiences involve ice or water, a substance which, like Ivy's *self*, is protean. Although in the form of frost, water can kill, this substance can also encompass awesome beauty and light and, at times, can work to create a transcendent moment for Ivy. In contrast to her letters about formal religion, Ivy thusly describes a winter morning in a letter to Mrs. Brown:

> *[I]t [the ice] was so pretty that it like to have took my breth away...It was like I looked out on the whole world and I culd see for miles, off down the mountain here, but it was new. The whole world was new, and it was like I was the onliest person that had ever looked upon it, and it was mine. It belonged to me...My breth hung like clouds in the air and the sun come up then, it like to have blinded me.* (18)

Rather than allowing her spirituality to be molded into a form acceptable to the Christian God as Ivy understands him, here she "authorizes" her own transcendent experience.

Hill, in her argument that Ivy reconciles the sexual with the sacred, the creative with the maternal, suggests that she does so through a calling up of forgotten archetypes. She argues that Smith calls up images of Lilith, Mary Magdalene, and a synthesized Aphrodite and Demeter and then uses them to undermine those archetypes embraced and imposed by the patriarchal structure, like that of the legendary King Arthur:

> *Smith's imagination, and particularly her linguistic sense, restlessly returns to root sources of cultural constructs that have to do with white/black, female/male, and human/animal. If the name Arthur [Ivy's father's name] does come from perhaps an animal and perhaps a female deity (the postulated root of Celtic arto- yields "bear" and "Arthur"), then all the boundaries drawn by patriarchal constructions—reified in heroic tales of such semi-historical figures as Arthur—are breached.* (114)

Hill builds a convincing argument that these feminine archetypes, defined as unorthodox by the patriarchy, subvert the power and the control of the "orthodox" ones. Smith does employ figures not only from Biblical scripture but also from mountain/Celtic sources. Yet, Hill's emphasis on the archetype as an inescapable source of truth is troubling. She even goes so far as to suggest "that one reason for this [the reconciliation of the father and the animal through Whitebear Whittington] feels so satisfying is that Smith's unerring linguistic sense has rediscovered an ancient connection" (114). The move that Hill lauds seems a positive one since it provides a way for Ivy to reconcile her creative self with her maternal self, something that is not often possible within the patriarchal system, yet Hill's suggestion is that the problem lies in society's ignoring the archetype of the "powerful goddess figures" (xix). This suggestion seems to imply that humanity would not have the problem of an oppressive power structure if we could only recognize and acknowledge all, rather than only some, of the archetypes etched into our psyches. This implication also necessarily carries with it the notion of a fixed and universal truth, inherent in the ancient archetypes.

However, what Ivy does is more complex than simply rejecting patriarchal Christian myths and embracing those of the powerful goddess. She juxtaposes letters, some of which express regard for the Christian God as an authority to be feared, others which position Him as an adversary whom she has the power to reject, and still others which dismiss Him as a relevant conduit to transcendent experience. In combining these conflicting positions, she constructs a world in which she is not enclosed within the fixed system of traditional Christian ideology, or the fixed systems predicated by the notion of a universal psyche, but in which she has the freedom to experience a variety of spiritually transcendent moments. Ivy's spiritual perspective encompasses many things, even contradictory notions. This interpretation more fully considers the challenges presented by the individual in the contemporary South and the contemporary world, the difficulty of preserving the elements of one's past as a factor in identity and consciousness while acknowledging that those elements are not entirely knowable or fixed and that they are often oppressive.

Just as Ivy's letters reveal a spiritually complex world, they also reflect a world in which the role of the patriarchal structure is inadequate and often burdensome, though it is a strong and inevitable component in Ivy's experience. In her family, consistent with her Southern culture, the land is handed from male to male so that even though Ivy and her mother and sisters work the land to survive, it is assumed that her oldest brother Victor will inherit it (36). Similarly, Mister Castle, Ivy's appropriately named maternal grandfather, decided that since his daughter Maude was marrying without his approval, she would be cut off from all of the family wealth and would be deprived of any of the advantages of the ties with her family. Yet, when Ivy's mother dies, rather than allowing her to be buried next to her husband on Blue Star Mountain, her father claims her body, taking it back to Rich Valley to be placed in his family's burial plot. Maude's own children have no say in the matter. Ivy's sister Beulah later experiences the oppressive nature of patriarchal society, as well. After the birth of her second son, she asks Doctor Gray to do something to "fix" her so she will not have any more children. Ivy writes, "[B]ut he said No mam, I can not in a definite Northern voice" (129). It is worth noting here, that although Southern culture is often criticized for its clinging to patriarchal social structures, Doctor Gray is portrayed as a Northerner, reminding us that patriarchal structures have dominated not only in regional cultures, but also nationally and internationally. Regardless, Smith investigates the restrictions that are brought to bear on the women of this Southern, Appalachian family and Ivy's complex perspective on them.

As with her treatment of religion, throughout the novel Ivy expresses mixed reactions to the patriarchy. Her assumptions include, in some letters, a resignation to the fact that being male is an advantage in such a society; for example, she does not question, as a young girl, Victor's inheritance of the land. On the other hand, there are some letters in which she expresses regret about the discrepancy in opportunities available to the sexes. After watching the boys and men on the river preparing to take the logs down to Kentucky, the teenaged Ivy writes to Silvaney, "I wuld give anything to be one of them boys and ride the rafts down to Kentucky on the great spring tide!…I will say I have even thoght of waring jeans and a boys shirt and shoes and trying to sneak along, but Momma and Geneva wuld have a fit" (86). She admits that she could never get away with such a scheme since her body has begun to change from a girl's to that of a woman. Inhibited by her blossoming and the resulting stares of the men and boys of Majestic, she expresses her wish to be small and plain like Jane Eyre, but "[i]nstead I am getting a bosom like Beulah, this is what they star at threw my dress." (80) Thus ends her hope of a river adventure akin to Huck Finn's.

However, in contrast to this attitude of unhappy submission, Ivy also writes letters expressing her refusal to accept the demands of patriarchal forces. She refuses to marry her first boyfriend, Lonnie Rash, even though it is known she has had sex with him, becoming, as she puts it, "ruint" (109). Instead, when she finds she is pregnant as a result of the affair, she moves to the company town at Diamond Fork mines to live with Beulah and her husband Curtis. Later on at Diamond Fork, when Oakley Fox, whom she has known all her life, claims that she is "his girl" and thus that she should not be seeing the wealthy Franklin Ransom, she writes her sister Ethel about the incident, stating that the idea is ridiculous because "I am not anybody's girl, Ethel" (154). Even though she sees Ransom and enjoys his attention and his wild lifestyle, she does not allow him to control her either. She notes with indignation his attempts to dominate her: "'Stick with me and I'll take you places,' Franklin said. 'You're my baby'" (166). After the affair goes sour, Ivy writes Silvaney of their last time together, explaining that he urged her to let him take her to Memphis, and when she refused, he attempted to rape her. She tells Silvaney that she stopped him, however, by asserting herself: "<u>I am not your baby</u>, I said" (166).

Finally, after her marriage to Oakley, Ivy and her family—not Victor—take over the Rowe family house and land and begin to farm again. After coming back from the war, Victor has gone into business in town and is not interested in farming the family's land. And it is Ivy, after all, who took to heart her father's words spoken to her as a child: "<u>Farming is pretty work</u>" (35). Ivy's relationship to the land is not defined fully by law or capitalism, as Victor's might have been—her very identity is linked to Sugar Fork, her connection to the place dynamic and respectful. Late in the novel, an aging Ivy expresses to her son Danny Ray her determination to preserve the land from a mining company that purchased the mineral rights from her mother years ago. Ivy describes the incident in which she stood pointing "Oakley's old thirty-ought-six" (307) at the man on the bulldozer who had come representing the Peabody Coal Company. In spite of the contract her mother signed with the company out of desperation for money, Ivy will not allow destruction of the place. In addition to breaking patriarchal convention by taking over the family farm, Ivy refuses to bend to the will of capitalistic forces, in the form of the Peabody Coal Company. Unlike her submission to the codes forbidding girls to ride down the river on the logs, she takes up agency to retain her ownership over not only her family's farm, but also the mineral rights, and thus to preserve the integrity of the land.

    Not only does Ivy resist domination by the patriarchal system, but her letters exhibit a blurring of the very boundaries of gender. She describes positively women who exhibit "masculine" characteristics, such as "Momma [who] stood too with her face as hard as a man's face" (41) and her friend Molly who has "a square hand like a man's" (304), as well as men who appear "feminine," like Revel who once dressed up like an old woman to avoid the wrath of the sheriff (39). Even more significantly, Ivy describes in a letter to Silvaney her merging with Honey Breeding during her affair with the bee man, whose name surely suggests gender blurring. Here, Ivy paints a picture in which the boundaries of their genders have no meaning: "But I am as big and strong as he is, and I toppled him into the starry flowers where we laid face to face and leg to leg and toe to toe. He is just the same size as me. In fact I think he *is* me, and I am him, and it will be so forever and ever" (230). Interestingly, Honey's existence seems to depend upon Ivy in a way that suggests that he *is* Ivy, that he is a figure constructed from her imagination. She writes again to Silvaney, "Sometimes now when I think of Honey Breeding, it is almost like I made him up out of thin air because I needed him so bad" (247). Yet, something happens to Ivy in recognizing this part of herself that is Honey Breeding. She is surprised at herself for finally making it up to the top of Blue Star Mountain at the urging of Honey, but "[a]ll of a sudden, I thought, I could of climbed up here by myself, anytime! But I had not. I remembered as girls how you and me would beg to go hunting on the mountain, Silvaney, but they said, That is for boys…And I had got up there myself at long last with a man it is true, but not a man like any I had ever seen before in all my life" (232-233). In some way, rather than requiring the companionship of a man to climb to this height, Ivy has accomplished it by acquiring (and writing) herself as a companion, a self who is not limited to one gender but encompasses both.

    Ivy's self is protean in that it is constantly moving, reshaping, impossible to define by any one particular ideology. Rather than rejecting the ideologies that have threatened to trap her and formulating her own position from which to battle them, Ivy shifts in and out of those oppressive ideologies, one moment employing their terms to define her circumstances and the next moment undermining those very terms. One can see the increasing complexity of that identity by looking at Ivy's initial definition of her self to her pen pal Hanneke: "I am a girl 12 years old very pretty I have very long hair and eight brothers and sisters and my Mother and my Father, he is ill...I want to be a famous writer when I grow up, I will write of Love" (14). This description compared with the sense of her self that Ivy conveys in her last letter reveals that, as Ivy has written, she *has* "been so many people": "I used to think I would be a writer. I thought then I would write of love (Ha!) but how little we know, we spend our years as a tale that is told I have spent my years so. I never became a writer atall. Instead I have loved, and loved, and loved. I am fair wore out with it" (316). Interestingly, this letter is written as the last snow is melting, and although Ivy writes that "[t]he Ice Queen walks in beauty like the night of shooting stars and cloudless skies," she also says that "[t]he ice is melting" (316). Like Ivy's identity, made up of multiple selves, the snow melts: "[A] hundred little rivers [run] down the yard and all of them shining" (315).

As I have suggested throughout this chapter, the role of the letter in creating Ivy's protean self is all-important. First, as we have seen, it serves as a form in which Ivy can mirror a fluctuating and fragmented self. Second, Ivy's letters are necessary for her reconciliation of the creative with the maternal. Without them she most likely would have failed to encompass the creative self even in her multifaceted identity since her choice of the maternal makes her life as a "real" writer—autonomous and "educated" in the conventional sense of the word—impossible. This reconciliation of Ivy's many identities, however, must be seen as *integrated* and not *unified*. MacKethan states that characters like Ivy "achieve reintegrations which restore their worlds, inner and outer, to wholeness" (*Daughters of Time* 102). However, Ivy does not achieve this reintegration at the expense of the deconstruction of the forces that bear on her. She does not become what we would call a unified character. Even at the end of *Fair and Tender Ladies*, Ivy considers herself a mother rather than a writer, regarding a writer as something other than a storyteller and letter-writer.

One way Ivy accomplishes this feat of *integration* is in her choice of Silvaney as a recipient of her letters. The letters to Silvaney include information that Ivy feels she can send only to this sister who for a short time is in a home for the mentally ill and who, not long after being sent to the home, dies in a fire. While all of her letters serve as a way for Ivy to see and create her self, her letters to Silvaney are particularly important. Often she can relate to this ever-absent sister what she cannot tell anyone else:

> *O Silvaney,*
> *I feel I am bursting with news but I can not tell it to a sole, I have no one to talk to... Oh Silvaney my love and my hart, I can*

talk to you for you do not understand, I can write you this letter too and tell you all the deep things, the things in my hart...And it is like you are a part of me Silvaney, in some way. So I can tell you things I would not tell another sole. (96-7)

Having established Silvaney as the recipient of her most intimate letters, Ivy has created a self that can exist in her otherwise harshly practical life. As MacKethan states, "Silvaney exists forcefully for Ivy as the part of herself who listens but has no voice, who lives on stories but has none of her own. In writing to Silvaney, Ivy connects voice and listener, teller and tale; in making these connections, she creates a sustaining, integrated image of herself" (*Daughters of Time* 108).[3] In establishing Silvaney as a character from a story in her imagination and as a silent listener, Ivy's letters to Silvaney not only allow Ivy to exist as Silvaney, "running wildly, silently, through the hills" (MacKethan, *Daughters of Time* 107), but also mirror a part of herself that she could not otherwise see, the self who has refused to give herself to God and has responded with excitement to her teacher Miss Torrington's kiss.

A third effect of Ivy's letter-writing is that her stories remain *one person's* stories rather than becoming what would conventionally be called "history." Consistent with her deconstruction of the ideologies that impinge upon her, Ivy keeps from recapitulating this impingement by refusing to set into ideological order the story of her own life. While she generalizes from time to time during the telling of her stories (e.g., "we have all got a true nature and we cant hide it" [280]), she will not pose as an authority through whom others may learn the "truth" of human existence. This position is well explained by Smith's description of her own attitude: "You just have to take all these accounts of the way it was and just kind of hope for the best, which is why I would really always prefer to call what I do *fiction*...I think I have a better chance of being *true* if I call it *fiction*" (Qtd. in Parrish, "Ghostland" 44). Ivy's decision to burn the letters she has written to Silvaney and kept in a chest for many years illustrates this necessary avoidance of producing "The Word," a permanent and imposing construct. In a letter to the worried Joli, Ivy explains that even though she has written to Silvaney all these years, she has not been confused—she has known her sister was dead, but she emphasizes that the *writing* of those

---

3 Kearns states, "Silvaney is a conduit to carry away madness, the sacrificial figure who remains necessary for the female artist to survive. Thus exercised, Ivy may accommodate the world without going mad, unlike her mother Maude" (190). Ivy needs a way, however, not only to objectify this self but also to encompass it in her overall identity. Through her letters, Ivy is able to hold on to this self rather than losing it entirely, viewing it in the mirror of her letters and absorbing it into her identity.

letters has been important to her. Nevertheless, she subsequently describes with delight her burning of these letters: "With every one I burned, my soul grew lighter, lighter as if it rose too with the smoke" (314). That the burning of the letters is so liberating for Ivy indicates that the existence of the letters, in which Ivy has continually defined her life, her *self*, is a potential trap. The burning of the letters frees her from the written definition of that self.

How does Ivy attain this protean state without becoming empty of substance, as does *Black Mountain Breakdown*'s Crystal Spangler? It is in *Fair and Tender Ladies* that Smith presents her first fully developed artist-protagonist. Although Ivy states that she never became a writer, we understand that she did. Her art is a unique one—she is a writer, but rather than becoming the novelist she wanted to be, she has written a body of letters, the whole of which connects her both to herself and to those she cares about. The letters to Silvaney are crucial in providing Ivy with a way to acknowledge all aspects of herself, some of which would not be acceptable to her community at large. But through these missives, Ivy also sustains her communications with Ethel, Joli, Molly, and many others. Like *Family Linen*'s Candy, Ivy's art has given her a medium for pursuing beauty and understanding, but also for developing and sustaining intimacy with her community.

Through her letters, Ivy has both reconciled and "unfixed" herself. In the act of writing, she has engaged in the creativity that by the standards of the patriarchal system might have been denied her in her choice of motherhood. In these letters, however, she represents a self that is multifaceted, that is "so many people." Her protean identity allows her to subvert the limiting definitions imposed by dominant ideologies and to move fluidly between, into, and beyond the boundaries of these ideologies. Thus through her art, Ivy demonstrates Smith's consideration of a positive postmodern state, one which can acknowledge the past but not be suffocated by it, one which enables individuality but also fosters community.

Chapter Six
"It was like I was *right there*": Primary Experience and the Role of Memory in *The Devil's Dream*

When Katie Cocker, of *The Devil's Dream*, begins to come full circle back to her roots after a circuitous journey through the music industry, it is a profound moment. Her return is provoked by the taste of her breakfast at Nashville's Loveless Café, where she sits with Ralph Handy, a fellow musician who once advised her to "[j]ust keep it country" (265). Sitting across the table from Ralph, talking over old times, Katie is unexpectedly pulled into her past in an almost physical way:

> *This was the kind of breakfast they used to serve up at Lucie and R.C.'s. I had not tasted sorghum molasses in years, but one taste of it made so many memories come flooding back. I told Ralph all about the molasses stir-offs we used to have up on Grassy Branch when Grandaddy Durwood was still alive, and how folks would come from far and near, and how good that molasses tasted when you dipped it up out of the stirring trough on a little piece of cane, and how the notes from R.C.'s banjo rang out in the still cold air. And all of a sudden I could see it—see the great fire and the full moon, it was like I was right there.* (287)

This experience, related to the reader by Katie herself, is significant for two reasons: 1) It illustrates the rewards, in Smith's fiction, of engaging in primary experience rather than secondary experience, and 2) it indicates, perhaps surprisingly, that *memory* can function as a primary rather than a secondary experience. With the above passage, and *The Devil's Dream* as a whole, Smith develops the assertion that as a potential link to authentic personal history, memory is crucial to the achievement of a meaningful sense of self.

I Have Been So Many People

In clarifying the meaning of the terms *primary* and *secondary* in this context, the philosophical distinctions made by D.T. Suzuki and Thomas Merton are helpful. Suzuki, explaining the problems inherent in dependence on the intellect as a mediator between the ego and the fact of the world, uses the term "personal experience" to express the concept I have been referring to as primary experience: "By personal experience it is meant to get at the fact at first hand and not through any intermediary, whatever this may be. Its favorite analogy is: to point at the moon a finger is needed, but woe to those who take the finger for the moon" (8). Merton, in his examination of Zen, relying heavily on Suzuki in his discussion, further elaborates on the distinction. He contrasts the Zen mind to the Cartesian consciousness:

> *It [the Zen mind] starts not from the thinking and self-aware subject but from Being, ontologically seen to be beyond and prior to the subject-object division. Underlying the subjective experience of the individual self there is an immediate experience of Being...It has none of the split and alienation that occurs when the subject becomes aware of itself as a quasi-object. The consciousness of Being...is an immediate experience that goes beyond reflective awareness.* (23-24)

The "Zen mind" is useful in examining Smith's 1992 novel since her interest in the post-Cartesian consciousness is developed so strikingly here. As Smith has examined in previous novels the inadequacy of traditional Southern historical ideologies to equip the postmodern subject for authentic life experiences, she offers in *The Devil's Dream* a way to connect with history in a more meaningful way, suggesting, as she does in *Family Linen*, that in spite of its pitfalls, the connection is indeed important. Unlike traditional "histories," which tend to serve as buffers against raw experience, memory, when experienced as Katie's in the passage above, enables the subject to both *feel* (as *On Agate Hill*'s Molly Petree desires to do) and connect with the past.

I do not frame this discussion with Zen principles to insinuate that Smith is a closet Zen Buddhist. Rather, I assert that there is significant similarity between Zen principles of non-mediated experience and Smith's treatment of the subject, and the comparison illuminates our understanding of Smith's theme in *The Devil's Dream*: For Smith, characters who alienate themselves from the realities of life around them, like *Oral History*'s Richard Burlage who engages in almost all events in a distinctly secondary way, seem to miss out on meaningful existence, while conversely there is a richness to the lives of those who, like Katie Cocker, participate in life more directly, or in a primary way. Smith's portrayal of these characters' experiences is consistent with the Buddhist principles explained by Suzuki and Merton. Unlike for Zen Buddhists, however, in Smith's fiction, primary experience does not negate a relationship with history—quite the contrary, in fact. Previous chapters in this study explore the contrasting problems of either clinging too tightly to an imagined and oversimplified history ("tak[ing] the finger for the moon") or denying one's history altogether. Both have negative effects on the developing self, restricting identity in a way that prevents us from full and meaningful experience and sometimes even healthy function. This chapter considers that while history itself is a construct, irrelevant and sometimes dangerous when taken as simple reality, its role is yet crucial as a factor of identity. Further, for Smith, *memory* can be a powerful primary experience, serving as a conduit to relevant relationship with the past and thus fostering authentic identity.

For Smith, life is more rewarding for those who engage in the risky enterprise of primary experience (including memory) than for those who insulate themselves against it. This distinction holds true in *The Devil's Dream*. Lizzie Bailey is, perhaps, the most obvious example of the result of alienating oneself by retreating to a state of strictly secondary experience. Daughter of Nonnie Hulett, who has abandoned her family for a medicine show man, Lizzie feels it is her fault her mother left home:

> [A]fter all, I was the one who wanted to go to the medicine show!...So I have done the very best I can, dedicated to erasing some of the harm done by those who run loose in the world... those who are messy and heedless, prisoners of their passions, unmindful of all others save themselves. I will not be like that, I had told myself over and over, and so I had worked my little fingers to the bone. (91)

Having suffered many of the difficulties common to women in her Appalachian culture, Lizzie desires a way out, a world of order as opposed to a "messy," poverty-stricken one: Upon her introduction to Miss Covington, she is immediately attracted to the independent life of the nurse. She sees all around her the fallout of engaging "too readily" in life: "The very notion of love terrified me...To me, 'falling in love' was like falling into death" (97). After years of serving as a surrogate mother to her sister Sally, Lizzie jumps at the opportunity to go with Miss Covington to pursue the life of a nurse. "I shan't forget my first glimpse of Miss Covington as she rounded the bend of Grassy Branch on her little mare, wearing a gray cloak over her gray split skirt, the little white cap perched firmly atop her pale blond hair, which she wore in a careful bun at the nape of her neck" (93). Lizzie's vision of Miss Covington emphasizes the controlled life she imagines the nurse to lead, her clothes and hair orderly, simple, and clean. Lizzie imagines the nurse's life as an escape from her own life of toil and volatility. She recognizes that "there was something in me, I see now, which needed Miss Covington—which *craved* her" (94). Years later, she expresses a certain amount of satisfaction with the life she has found, away from the expectations of her family and mountain community. Yet, there is an uneasiness in her voice as she tells of her plans to go to France to serve during World War I as a nurse: "How can it be, I wonder, that the closer our date of departure draws, the less I am able to even *imagine* France, and the more I find myself traveling back through time and circumstance, back to my Virginia childhood?" (90). Significantly, until now, she has disassociated herself from the memories of her childhood to avoid the pain they bring, and this insulation is one example of the life she has cultivated—a life of only secondary experience, in which she experiences everything as it has been intellectually packaged (and made innocuous in the process). In this moment of transition, as she allows herself to remember, she describes a split between the Lizzie she sees as her self *now* and the Lizzie she was *then*. It is only in her remembering, in her writing her memory, that she reaches a point of recognition, that "I am able to see myself as even vaguely the same person I was then: *only now* am I able to do this" (106). Understandably, Rebecca Smith sees Lizzie's escape from her mountain culture as a positive move:

> *Though her turning down marriage to a doctor and her preference for an ordered life free of the "clotted, messy, tangled" web of family and sexual connections...may make Lizzie seem sterile and inept, her successful work as a nurse and her eventual death in France during the war characterize her as one of Smith's first women whose work makes a difference in the public world.* ("Writing, Singing and Hearing a New Voice: Lee Smith's *The Devil's Dream*" 54-55)

Certainly, as readers, we want Lizzie to find a space for herself, one which she can fill without obliterating her own desires. We can certainly see her independence as a nurse as a positive alternative to Crystal Spangler's catatonia, a result of her always accommodating others' expectations. Yet, Lizzie's acknowledgment that the possibility of being fundamentally affected by life produces in her "the strangest, most unpleasant sensations—light-headedness, nausea, shortness of breath" (106)—leads us to the conclusion that her escape, her "victory," is ultimately an empty one. She describes her sterile, though safe and predictable, existence: "So on the whole, I prefer a more professional involvement with the human race. I prefer situations I can at least hope to control: a bone to set, an arm to bandage, a cut to stitch up, a set of instruments to sterilize" (107). As a reader, it is difficult to ignore that by walling herself into a life characterized by secondary experience only, Lizzie becomes isolated and joyless. And reflective of that state, her identity is only partial, comprised primarily of her function as a sterile instrument herself, cut off from historical and cultural context.

Similarly, Alice Cocker, though a wife and a mother, mediates her experience through her religion, relegating the challenging details of her life to the secondary realm as Lizzie does. She is able to shield herself, from her husband's alcoholism, from her children, and from her own desire, by concentrating on the spirit rather than the flesh. Though her "faith" may in some ways be said to preserve her through a difficult life, we are led to believe that, even by the time she is an old woman, she has not really *lived* at all. When she is presented with her own great grandchild, Katie's grandchild, she seems almost afraid to touch the baby: "'No,' she says, looking away, drawing up her face till she looks like a dried apple doll" (307). It is not until she becomes caught up in the experience of the reunion with her family that she says, "Well, why don't you hand that baby on over here after all?" (310)

Although Zinnia Hulett gives in to the expectations of her Appalachian community, like Lizzie and Alice, she is still alienated from meaningful experience by her denial of her own desires. Her role as wife and mother is performed only as a surrogate and vicariously through her sister, Nonnie. It is mediated, thus secondary. Though her community may be blamed for their complicity in this alienation, relegating her to the margins of their activities, Zinnia seems willingly to remain an outsider. She fears, and even disdains, the results of plunging into experience, as is reflected in her criticism of Nonnie's constant singing: "I could not carry a tune in a bucket myself, and don't give a damn to. For what good does it do you in the end? What good did it do Nonnie?" (53). She condemns Nonnie for her passionate behavior, though the undertone of all her criticism rings clearly of bitterness and jealousy.

Conversely, while Appalachian life can be restrictive, physically difficult, and even violent at times, especially for women, the characters who are able to take their chances with these dangers and confront life "head on" reap a kind of rich experience that the likes of Lizzie do not. For example, while she is still but a girl, Nonnie Hulett runs headlong into an affair with Jake Toney, a "Melungeon," a man of mixed heritage who is shunned by Nonnie's community, and as a result, she ends up cast out of this community herself. Her feelings about Jake reveal much about her: She has little control over her passion for him. When he comes to court her and her father sends him away, she restrains herself for a moment, yet according to her sister Zinnia, "as soon as he was gone, she just threw herself on Daddy like a wildcat from Hell, crying and clawing at his eyes and hitting at him" (56-7). When it is discovered that she is pregnant, she is shipped off to a distant community to marry Ezekiel Bailey, a man of limited intellectual capability, because as preacher Cisco Estep says to Nonnie's father, "Well now, Claude, think about it...If she tried to come to meeting in her condition and unwed, I'd be forced to church her, as ye know. And around here, everybody knows who she is and what she done, and won't nobody take a Melungeon's leavings around here neither" (59). Reflecting the split in Lizzie's identity, between the Appalachian girl and the more sophisticated and "worldly" nurse, Nonnie leaves behind, for awhile, her impulsive girlish self to become a wife and mother. Nonnie is full with emotion over her life, though. Difficult as her new lot is, she does not lapse into secondary experience to buffer the pain, as Lizzie does. For example, when she finally loses the sense of satisfaction with her role as mother and wife, she becomes conscious of the split that has occurred, eclipsing a crucial part of her, and she runs off with "Dr." Harry Sharp of the medicine show to recapture that self. Her experience as she becomes caught up in the show is primary, both in her complete engagement in the show itself and in her reconnection with the repressed self from the past through memory: "It was at this precise moment that Nonnie felt it stealing over her, that feeling from long ago, that quivering mixture of excitement and longing and dread which meant *Now. Right now*" (75). The music of her childhood, which she finds again at the medicine show, works to bring the past into the present and to produce the state of unmediated being, or primary experience, which overcomes Nonnie. When Harry asks for audience participation, she walks up to the stage, almost involuntarily, and says she'd like to sing:

> *And so Nonnie sang the song she'd sung when her daddy put her up on the counter as a young girl, all those years ago, her high pretty voice trilling on the last line, "And she never sings cuckoo till the spring of the year," and for a minute, she was that little girl again, so silly and so good.* (76)

In spite of its implications as to the sadly limited options for Appalachian women, the moment of Nonnie's unmediated reliving of her past reveals her willingness to embrace experience firsthand, including this relevant and intense event of her personal history. Her identity, reflected in this moment, encompasses the present and the past, convergent factors of Nonnie's self. In Nonnie's choice of primary over secondary experience at this point, we sense that, unlike Lizzie and Zinnia, she will *not* go with Harry, regardless of the risk or of her obligations as a "good" girl or as a mother and wife. Though her decision leads her eventually to her death by burning, Nonnie has never been afraid of fire. She exemplifies Smith's assertion that only those with a dynamic sense of self, including a relevant connection to their past, can engage meaningfully in life.

Like Nonnie, Lucie Queen seems reckless in her willingness to take chances, but Smith implies that her way of interfacing with life is richer than that of more careful characters. The young Lizzie, while still at home, worries over Lucie's naïve adventurousness. She even tries to discourage Lucie from marrying Lizzie's brother R.C.: "'Lucie, it is not a game. It is not a play-party,' I said between my teeth. 'You are fooling with fire'" (98). Yet, Lucie will not accept the notion that she should distance herself from the fire. Rather, like Nonnie, she is ready to walk headlong into it. Her relationship with R.C. is a passionate one. According to Lizzie, "They looked flushed, intoxicated, as if they'd contracted a fever" (99). But even Lizzie admits that "[t]hey did seem happy, R.C. and Lucie" (101). Lucie's tendency toward primary experience carries over into her music, as well, even as she and R.C. begin to perform publicly, playing songs connected closely to their mountain life, songs like "The Devil's Dream," "Shall We Gather at the River," and "Shady Grove" (104). Lizzie states, "The secret of their success, I feel, was Lucie—so unaffected, so guileless" (103). They meet with praise from the people in town where they play "at square dances, play-parties…house raisings, bean stringings, too" (103). Significantly, these locals experience *as primary* Lucie's and R.C's songs, which are the community's own, expressing their own stories and sensibilities. These songs do not, perhaps, fit neatly into the category of *memory*, yet they are a mechanism for reflecting the culture that produces them. Art rather than documentary, they still provide a conduit by which people can connect with their own pasts, their own cultural contexts.

This point is made clear in light of the fact that when taken out of context, the songs lose their ability to function as such conduits. During the recording session at Bristol, Lucie seems to recognize the transition that occurs as the music becomes commercial, transformed from an agent of primary experience for those whose lives are expressed and shaped by it to a mediator for those who adopt it artificially, as a kind of secondary experience (the "finger" rather than the "moon"). Nancy C. Parrish characterizes the problem in her discussion of a similar transformation, that of the Cantrell mountain homeplace in *Oral History* into a prop for the theme park, Ghostland:

> *Although the burial ground and homeplace are preserved "untouched," they exist out of context as folklore backdrops for amusement. Even being described as "untouched" is an inaccuracy, because the burial ground and homeplace cannot have the same meaning they had before they were situated in the context of the theme park. This image confirms the linguist's sense that the context can indeed determine meaning.* ("Ghostland" 45)

Lucie seems to understand the seriousness of this discrepancy: "Head down under the pretext of tending to the baby, Lucie cries softly. For it seems to her that they have just given up something precious by singing these songs here to these strangers, and she feels a sudden terrible sense of loss" (124). Though Lucie may not intellectualize over the transition, her response indicates a recognition that the music will not have the same meaning to an audience that does not relate to it directly but rather accepts it as a mediator to experience.

While Smith clearly privileges primary experience, she treats memory, perhaps surprisingly, as crucial to a full understanding of self, as is exemplified by Nonnie's realization of her self in singing the "Cuckoo Song" at the medicine show. In contrast, although her husband Ezekiel Bailey is clearly an affectionate and good man, there is something missing with him—he lacks the capacity for memory. Like Devere of *Black Mountain Breakdown*, Ezekiel exists only in the present: "[M]ost times he could not remember his mother, so that when Great-aunt Edith told him she was dead, that first summer after they took him away from her, it meant nothing, nothing to him at all" (35). Though Ezekiel is a hardworking and generally contented man—engaging *wholly* in primary experiences, such as sex, religious ecstasies, and physical labor—he is virtually incapable of meaningful experience or a sense of self because of his inability to remember. Even after grieving over the loss of his father-figure, Preacher Stump, and deciding to name his son Reese after his mentor, Ezekiel's feelings of loss fade quickly, "as most feelings did with him, and by the time R.C. was born, he had almost forgotten the old man" (66).

This emphasis on the importance of memory is where Smith diverges from the path of Zen. Alan W. Watts, in *The Way of Zen*, explains that a crucial precept of Zen is the recognition that the past is not real: "Man's identification with his idea of himself gives him a specious and precarious sense of permanence. For this idea is relatively fixed, being based upon carefully selected memories of his past, memories which have a preserved and fixed character" (122). It is this holding onto one's idea of oneself, comprising memories, among other things, that, according to Watts, keeps one from achieving spiritual liberation. Smith's fiction, consistent with Zen principles, docs suggest that constructing one's self from a fixed notion of history, shaped usually by socially determined values, is problematic. For Smith's characters, however, failure or inability to embrace the past can leave one without a meaningful sense of one's self.

Katie Cocker exemplifies the effects of both ends of this spectrum. Tired of the limits placed on her by her life in Grassy Branch with her mother Alice and her grandmother Tampa, she leaves to pursue a career in country music. In order to market herself for success, she accommodates the audiences' expectations by taking on first one identity and then another. As one of Mamma Rainette's Raindrops, she poses as a dumb country girl, pretending to be "just short of retarded" and wearing "straw hats and bloomers and big black clodhopper lace-up boots, our red-checkered dresses buttoned up wrong" (225). Later, when she allows Wayne Rickets to become her manager, she agrees to a new image: "He stuck me into a push-up bra and four-inch heels and the fanciest low-cut outfits you ever saw…At first I balked when he suggested the wig, but then I started wearing it too" (252). These "show images," shaped heavily by stereotypes of Southern culture, are followed by those of the "California singer" and the "good country woman" (299). These identities are imposed upon her by the producers and consumers engaged in imagining her (and their own) history and failing to do so accurately. Thus the false personae she takes up do not offer her or the audience an authentic connection to the folk culture that has produced the music she sings.

Finally, after surviving her daughter Annie May's polio, her own bout with alcoholism, and the tragic death of Ralph Handy, her long-time friend and in later years her husband, Katie finds herself alone and image-free. As Rebecca Smith suggests, "She learns to create her own image at the end of the novel, thus standing as a woman artist who finally overcomes the demands for conformity to a subjugated feminine role" ("Writing, Singing and Hearing a New Voice: Lee Smith's *The Devil's Dream*" 56). But, overcoming those demands has not alone brought her to the place she has carved out for herself at the end of the novel. She has first had to reach an understanding of herself in reconciliation with her past. Her new independence is coupled with Katie's decision that "you *can* go home again," a realization necessary to her moving beyond another stop-gap identity to a complete healing and a healthy sense of who she is.

When she decides to produce an album on her own, filled with songs from her past, it is a coming home. She calls the album *Shall We Gather at the River* and includes songs like "Down by Grassy Branch" and "Melungeon Man," songs that connect her with her youth, her family, and the mountains. Debbie Wesley notes that

> *In* The Devil's Dream, *Smith emphasizes that her artist must maintain not only her ties to her immediate family but also to the community from which she sprung [sic]...Katie has discovered that her family and her past have shaped her into the unique individual that she is, and now she has the confidence to unite with her family without feeling that they will overwhelm or smother her.* (100)

Interestingly, Katie finds religion, realizing that she has been avoiding the spiritual to keep from following in her mother Alice's footsteps. This recognition that she has been running from her own history and, in a spiritual sense, her self, is an important step in her growth. Yet, Katie's spirituality is different than her mother's. It is a more primary experience than Alice's religion seems to be. Katie opens herself to God in an ecstatic moment: "I could feel my pain rushing up from all over my body, feel the shock when it hit the air, feel it shatter and blow away, nothing but dust in the wind. Then I felt God come into me, right into me through the mouth, like a long cool drink of water" (298). This conversion does not cause Katie to separate herself from the "carnal world," as Alice's religion teaches its followers to do. Even after she rediscovers the spiritual, Katie continues to live a physical, sensual existence: "[D]on't get me wrong. I still know how to have a good time. I like to dance. I will take a drink from time to time. I like to have a date. There's nothing wrong with any of this. Billy Jack says that, above all, God does not want us to put ourselves under a bushel" (302). Katie is able to both embrace her past and continue to live life in a primary way, something many of Smith's characters are unable to accomplish. Wesley captures the fearlessness that Katie exhibits throughout her life—her tendency toward primary experience rather than secondary: "Following her heart rather than her head may get Katie into trouble, but it also brings her joy. Smith clearly admires individuals who act spontaneously and instinctively and yet have the courage to take full responsibility for their behavior" (98). Ultimately, Katie achieves a healthy sense of self because of her willingness to embrace her past as part of her direct interface with life in general.

The equation I have set up to describe Smith's advice for self-fulfillment is not as simple, of course, as may be implied by the discussion up to this point. The mental breakdown of Katie's cousin, Rose Annie, seems to result from her inability to embrace the *present*. Completely fragmented, she lives in constant attempt to recreate her passionate relationship with her cousin Johnny, to *recreate* the past. Lizzie's separation from her past, her self, seems an attempt to avoid the kind of consequences that Rose Annie experiences, the violence that the past can produce. How can we condemn her for that act of self-preservation? These seeming discrepancies are characteristic of Smith's fiction. Like the principles of Zen, she is a moving target. The very fact that we learn, in a secondary way, of Katie's primary experience in coming to terms with her past seems paradoxical. Smith is a fiction writer, which necessarily casts her in the role of constructing secondary realities and us in the role of consuming them.

It is this tendency to avoid ultimately simple explanations that makes Smith's fiction so rich. It is also one of the reasons her work can be fruitfully regarded through the lens of Eastern thought. Smith appears to forward in her work the necessity of what Maxine Hong Kingston's woman warrior, Fa Mu Lan, is taught during her training: "I learned to make my mind large, as the universe is large, so that there is room for paradoxes" (29).

Even so, the reader generally comes away from Smith's stories with a high regard for what some would call "the real." Katie Cocker is an earthy, tough, and insightful character whom we like in a way that we simply cannot like Lizzie Bailey. She has suffered in ways that Lizzie has not, perhaps because she is naïve and even reckless, but she comes out ahead, according to the narrative, in that she is engaged with life, including her past, in a direct way, refusing mediation. Smith's fiction confirms, at least most of the time, what Merton suggests, that one crucial need of humankind is "liberation from...inordinate self-consciousness...monumental self-awareness...[and] obsession with self-affirmation, so that [we] may enjoy the freedom from concern that goes with being simply what [we are] and accepting things as they are in order to work with them as [we] can" (31). While Smith may seem to challenge the tenet of Zen that suggests that one must cease identifying oneself as the sum total of one's past, she affirms through her emphasis on the importance of memory as a vital connection to the past that Merton's notion of self-acceptance is achievable only inasmuch as one is willing to accept, and even embrace, the past. And this embrace is most relevantly and usefully achieved through the primary experience of memory.

Chapter Seven
And the Word Was God: Narrative Negotiation of the Spirit/Flesh Split in *Saving Grace*

Lee Smith's novel *Saving Grace* has provoked debate among readers and critics since its publication in 1995. The story traces the life of Florida Grace Shepherd from her arrival as a child at Scrabble Creek in the North Carolina mountains, during her travels away from Scrabble Creek and into Tennessee, through her encounters with marriage, love, motherhood, spiritual struggle and sexual awakening, and back to her North Carolina homeplace at the end of the novel. Having left Scrabble Creek as a girl following the suicide of her mother, Gracie returns here as a middle-aged woman to the one place she ever felt "at home," looking to find her*self*, her place. She has come full circle. According to Linda J. Byrd-Cook, Grace experiences, upon this return, a new sense of her self, a kind of "salvation," in a reconciliation with her mother, with The Mother, as she handles live coals in her mother's ghostly presence, during a moment of spiritual ecstasy.[1] Yet, after this episode, Grace departs the house on her way to...where? Smith does not tell.

Grace's state of mind as she prepares to leave Scrabble Creek this time implies that she will not be back: "I stop for one last time to kneel by the icy rushing waters of Scrabble Creek. The sweetest sound I ever heard, it has stayed in my head all these years" (272). She begins to hear the cries of her dead infant son, buried years ago: "That baby is crying again, you know I left him outside crying in the dirty snow but I am coming now, I am really coming Jesus" (272).

This ending has spurred not only a variety of interpretations, but also a range of opinions regarding the literary merits of the novel. While scholars such as Dorothy Scura feel that *Saving Grace* is "perhaps, not up to the

---

[1] See Byrd-Cook's "Reconciliation with the Great Mother Goddess in Lee Smith's *Saving Grace*."

standard of *Oral History* or *Fair and Tender Ladies*" (as qtd. in Hall 90), others, like Jacqueline Doyle, defend the book as complex and daring, "a radically revised Christian spirituality that is open-ended, indeterminate, mysterious, and woman-centered" (3). It is true this novel certainly does not reach the level of impact that *Oral History* and *Fair and Tender Ladies* do. It is flawed, as it seems the believable options left for Grace at the end of the novel would not produce the kind of optimism with which she drives away from Scrabble Creek. The ending is problematic. In spite of this failure, however, the novel is yet provocative, complex, and powerful.

In previous novels, in her exploration of the *self* and the factors that shape it and destabilize it, Smith has examined ideals of Southern family and gentility, notions of history, and the roles art and memory can play in this process of identity-pursuit. While evangelical Christianity certainly rears its head in *Black Mountain Breakdown* and *Fair and Tender Ladies*, it is not until *Saving Grace* that Smith delves fully into an investigation of the profound influence of Southern Appalachian religion. As she traces its influence in the lives of this novel's characters, she characteristically examines its role in shaping adherents' identities as well as the ways that Gracie Shepherd resists its often stultifying effects.

Consistent in her resistance to pedantics, Smith does not, in this novel, offer a rigorous critique of the Christian ideology or even a clear condemnation of its charismatic practice in poor regions of Appalachia. But neither does the story's spiritual perspective match the orthodoxy of Flannery O'Connor's vision. Instead, as in all her fiction, Smith explores the issues at hand with story and leaves them essentially unresolved. In shaping this narrative, she employs the paradoxical language of dualistic, Protestant, Christian ideology. Significantly, however, while the ambiguous ending leaves these paradoxes unresolved, the contradictions manifested within Grace are reconciled, in a sense, by a language of fluidity. Smith's maintenance of religious paradox and simultaneous infusion of this language of fluidity allow for a narrative tension that renders a strange product, definitely flawed, yet undeniably captivating. Through this strategy, Smith illustrates a contemporary identity by which Grace both participates in her traditional cultural mythology and subverts its stultifying power.

Smith's portrayal of Protestant dualism, specifically of the Appalachian Holiness doctrine, begins with the linguistic distinction of the *carnal* from the *sacred*, the flesh from the spirit. Gracie voices her understanding of this structure when she describes herself to the reader: "They say I take after her [Grace's mother, Fannie], and I am proud of this, for she was as lovely as the day is long, in spirit as well as flesh. It isn't true, however.

I am and always have been contentious and ornery, full of fear and doubt in a family of believers" (3). Gracie's perspective on this bifurcation is undoubtedly formed, at least in part, by the language she has inherited for articulating the relationship between body and spirit.

Grace's knowledge about the carnal state, ironically, comes most directly from the example of her preacher/father, Virgil Shepherd. Always a man ruled by passion, Virgil not only is often "anointed" (39) with the Spirit, but also frequently engages in sexual affairs with women of his revival congregations. Grace's sister Evelyn reveals his philandering to Grace during their early days at Scrabble Creek: "'You know what I think is wrong with Daddy? You know what I think is wrong?...He does everything too much,' Evelyn said darkly. 'Whatever it is. If it's God or if it's a girlfriend, it don't matter. He does it too much'" (51). While Evelyn can see the overlap between the ostensibly spiritual and the physical in this case, Grace uses the dualistic terminology familiar to Southern Protestants to describe her father's indulgences. She says he has "backslid" (59) and "fell by the wayside" (122), reminding us of the traditional images of Christian spiritual journey as a difficult trek up a steep and narrow path.

The carnal, for Grace, is departure from this path. Grace herself is drawn to the earthly/carnal world from early adolescence on. She desires the comfortable and luxurious life that her friend Marie Royal exposes her to and is resentful that her life of Holiness excludes the possibilities of that material comfort. But after her mother's suicide, she gives in to "God's will," which for a while she equates with Virgil's will, and goes on the road with him to spread the Word. She states, "I was an instrument of Daddy, the way he was an instrument of God. I understood this, and bore it without complaint. I felt like it was my due some way, my duty" (121). Clearly, Grace has taken up a prescribed identity here in her role as her father's evangelical helpmate. However, she soon grows weary of her father's inconsistency, his tendency to disappear when he meets an attractive woman at meetings. After her father finally abandons her to abscond with the sensual and "crazy" congregation member, Carlean Combs, Grace sets her sights on the young minister Travis Word. Travis is committed to the traditional notion of the difficult spiritual journey based in the dualistic language and ideology of Grace's upbringing, and unlike her father, he is consistent in striving to stay on the path of righteousness. At least early on, Grace appreciates his commitment to the spirit over the flesh: "Travis Word was the first preacher I ever ran into that placed works above grace in order of importance. As a person even then searching for hard ground in a world of shifting sands, I liked this. I was real glad to hear it" (164). When Grace joyously consummates her marriage to the stern and solemn preacher, Travis regrets his passionate behavior and argues that "those who are in the flesh cannot please God" (188). The recollection of this Biblical verse provokes Travis to spend the rest of the night on his knees praying, trying to reconcile Grace and himself once more to God.

This reaction to physical passion is strange to Grace, but early in her marriage she can tolerate and even appreciate Travis's asceticism. During this phase of her life, she not only accepts the body/spirit dichotomy, but she is grateful for the stability it brings to her previously tumultuous life. As the years pass, though, Travis's simple way of seeing his relationship to God troubles Grace: "Travis believed that everything in life happened for a purpose and fell into the great scheme of God, but I did not. I was still prone to question and agonize" (202). Over time, Travis's insistence on respecting this dichotomy leaves Grace weary with the marriage and her role in this system.

Transcendence, the other aspect of the traditional flesh/spirit duality, is often defined in direct opposition to the carnal state. According to common Protestant precepts, if the desires of the flesh are sinful, then to deny oneself fulfillment of those desires is to glorify God. Yet, even as early as her wedding reception, Grace feels a pang of resentment when she sees that Garnet Keen has scripted on the wedding cake, "'Glory to God Amen'…instead of 'Travis and Gracie in Love,' which she had said she was going to write" (181). Grace's irritation is only fleeting, though, as she resigns herself to giving up most of her personal desires if she is to be a preacher's wife. At this early stage, she buys into the notion that the transcendent and the earthly are distinct from one another.

Throughout the novel, Grace's understanding of the sacred includes its manifestation in the physical world in the form of "signs and wonders" (6). Virgil is always looking for these signs and applies the term to events that seem too coincidental to have happened by way of natural law. He claims, "As it says in the good Bible, this world is not our home, we're only passing through" (9), and as a spiritual traveler, he looks for God's hand, when he is not distracted by women, to direct his path. He takes the explosion of the family car as such a sign. Grace relates:

> *"Bless Jesus" Daddy said, reaching his arms up in the air and bowing his white head to the will of God…"Bless Jesus…who has showed us by the sign of fire in his holy woods nine miles out of Waynesville, North Carolina, His plan for our life today, by freeing us from the things of this world and casting us wholly on his mercy, amen."*
>
> *…Daddy appeared beatified.* (9)

These "signs and wonders" are, for the Holiness people, evidence of the sacred beyond the fallen world. The Holiness congregation is aware of the spiritual realm as an eternal presence. When Virgil's ministry successfully converts nonbelievers, this conversion is referred to as bringing "souls... into glory" (122). The term "saved" has special significance in this context, implying that the convert has been reborn as a child of God and has, in the process, escaped the penalty of spiritual death. The Word of God itself, containing the divine revelation of God, is honored by these believers above all other written texts. Travis Word, aptly named, has forsaken other sources of learning in order to defer to God's Word. Gracie explains, "In fact, he could have gone to college scot-free due to the basketball, but after struggling with it all, he decided not to, as he felt himself in danger of placing basketball above God, and other books above the Bible" (167). The signs of the spiritual, the possibility of salvation from carnality and its consequences, and the existence of God's divine Word in the fleshly world confirm for Grace and her Holiness community that the spiritual can and must be distinguished from the physical. In spite of her own resistance to salvation at various points in the story, Grace believes in the flesh/spirit split, for the most part. Late in the novel, when she returns to see the decay that has occurred in the Scrabble Creek cabin, she says, "All returns to the earth, and the Spirit returns to God who gave it" (270).

If Christianity is, then, concerned with bridging the chasm between the dual realms of flesh and spirit, for Grace's Holiness Church community, the *ecstatic moment* is a fleeting reconciliation of the two. Smith has a particular interest in this moment: During this instant when the flesh is transcended temporarily, the soul stands on the threshold between earth and heaven. The most commonly acknowledged experience of this threshold is the passage from life to death. Thus, it is not surprising that, for Grace's Holiness community, the ecstatic spiritual experience is often a close call with death. Snakes and poison are the catalysts that Virgil Shepherd prefers. During snake handling sessions, the ordinary constraints of physical existence seem to be overcome, so that those involved can hold snakes and not be bitten and can even speak the language of God. According to Grace, on one such occasion, Virgil was

> *fully anointed and covered in snakes, which writhed around his arms and his body and popped out of the open front of his shirt...There were those present who will swear to this day that Daddy gave off light like the sun. Many were blessed that night to come forward and pluck the serpents off of Daddy and handle them too, and nobody was harmed in any way, and tongues of*

> *fire came down on several, including Carlton Duty who flung back his head and shouted out in the unknown language of the Lord for upwards of an hour.* (22-3)

Similar effects result from Virgil's drinking poison to spur the transcendent experience:

> *I [Grace] was present at Sunday morning meeting in the Jesus Name Church to see Daddy bring in the mason jar of water with Strychnine in it for the first time, and I saw him drink it with no ill effects, though an unbeliever grabbed it up to try it and then went screaming from the church house clutching his throat. Soon all the other saints were drinking poison too, whenever God moved on them to do it, and this was a regular feature of Daddy's ministry.* (50)

Grace's mother Fannie, and later Grace herself, experience the desired ecstasy by way of handling live coals. As a child, Grace is petrified when her mother engages in this strange ritual. Rather than pray during this event as her mother tells her to, Grace "was banging my head on the floor and saying 'I hate Jesus! I hate Jesus!' over and over in my mind, because he was burning my mother" (26). But her mother reassures her, "It was a pleasure in the Lord. Oh honey, I don't know how to explain it. It was just too good to explain" (26). For the Holiness worshippers, these ecstatic moments are crucial to transcendence of the flesh.

It would be easy to veer into mockery in portraying this dramatic and, by mainstream standards, strange practice of courting the ecstatic. But Smith does not allow herself to do so. Her fascination with the role of the ecstatic moment, a distinctly primary experience, drives the narrative in these passages and yields a respectful and rich impression. While most mainstream American Protestant churches would find these practices extreme, Smith's emphasis on the transcendent moment keeps our focus on the overarching question posed by Christian ideology: How do the fallen experience and connect with the spiritual? Or from a more secular and postmodern perspective, we might ask, how can one draw *meaning* from material existence? By emphasizing this central question (which, Smith implies, can be posed in a variety of ways), Smith leads us always back to the issue of Grace's identity, her formative self, and the role that this dualistic ideology plays in that formation.

Interestingly, the dichotomous structures of this belief system, a system which for centuries dominated the Western world and which has heavily influenced Southern culture, ultimately fall short for Grace. Even with the possibilities of transcendence that the church offers, and even though

I Have Been So Many People

Grace acknowledges that she has been given the "gift of discernment" (111), she resists the pressure to surrender completely to God. Her ability to embrace her religion fully is thwarted by four paradoxes: 1) love for God often pitted against love for fellow humans, 2) appreciation of earthly pleasure vs. devotion to the sacred, 3) fulfillment of duty vs. openness to spiritual ecstasy, and 4) justice vs. grace.

Grace experiences, early on, the dilemma of having to choose between utter devotion to God and the interests of family members whom she loves. When the illness of her little brother, Troy Lee, reaches a critical point, the family knows he is verging on death, yet as "believers" they leave his fate to God rather than taking him to a doctor. Joe Allen, Grace's oldest brother who has moved into town to work, decides he must act in order to save Troy Lee and comes to retrieve him and take him to the hospital. When Virgil catches Joe Allen in the process, he resorts to physical violence to stop his oldest son. As Fannie lifts the "big pickling crock" (63), ostensibly to aid Virgil in stopping Joe Allen, Gracie "reached up and grabbed it and sent it rolling out the door and off the porch. I put my arms around Mama from behind and held her tight, which was easy, for she was so slight and I was strong" (63). A few moments later, when Joe Allen has escaped with Troy Lee in his arms, Grace

> *let go of Mama, who turned around and looked at me in a tragic way. "Oh, Sissy, the devil has done claimed you for his own," she said, and rushed over to Daddy, who still sat there against the chair like he didn't know what was going on. Mama knelt down on the floor and covered his bloody face with kisses.* (63)

This dilemma of having to decide between physically saving their brother and waiting to see if God intends for him to live illustrates the "tests of faith" that challenge Grace.

Despite this momentary choice of allegiance to her brothers over her duty to God, however, Grace struggles for years afterward to resolve the issue for herself. After her mother commits suicide because of Virgil's unfaithfulness to her, Virgil orders Grace to leave Scrabble Creek and go on the road with him and his ministry. She hates to leave her dependent sister Billie Jean, who stays behind in North Carolina with Ruth and Carlton Duty. Nonetheless, she opts to go with Virgil to fulfill what she perceives as her duty. Grace does have some reservations buried beneath her faith. She is not surprised when her father abandons her in East Tennessee, and she keenly feels the sting of his neglect in favor of his "calling." She aptly comments on the contradiction: "Mama took good care of us, as good as she could. This was not true of Daddy, nor of Jesus either, as far as I could see" (3-4). Grace cannot make sense of her observation that people's devotion to one another often seems to clash with their allegiance to God.

Second, Grace negotiates the conflict between her appreciation of physical and emotional pleasure and an obligatory regard for the sacred, although she often gives in to the former. Her struggle results in the seemingly impossible *merging* of the two at times, as is revealed in her response to Ruth Duty's gift of cake. When the family car catches fire and leaves them stranded and hungry on the mountain road between Georgia and North Carolina, Ruth and Carlton Duty stop to offer help. Ruth brings out a coconut cake to feed them. Grace remembers the meal vividly: "In my whole life, I have never tasted anything to equal Mrs. Ruth Duty's coconut cake. Even today, it makes my mouth water just to think about it! I reckon we ate like we were starved, which we were…I thought we had died and gone to Heaven for sure" (8).

Similarly, Grace equates particular moments of motherhood with what she believes Heaven to be. She describes the days when her daughters were still just babies: "Each day seemed to stretch out full and golden, and last forever. I was the happiest I had ever been…Of course I was happy! I was living in a paradise" (193). In the most poignant merger of the physical with the sacred, Grace characterizes sexual pleasure as ecstatic, much like she has described the moments of spiritual transcendence she has witnessed and experienced. In the middle of a revival meeting to which Grace is responding with fervor and emotion, her dark and seductive half-brother Lamar takes advantage of her high emotional state and guides her out of the tent toward a sexual encounter, "and I went with him. Yes I did. I'll admit it. I was swept along, carried away in the general fever of that night. I did not feel that I was doing wrong either, even then" (106). Her affair with her half-brother seems somehow natural to her in the context of this ecstatic spirituality. Similarly, after her first tryst with Randy Newhouse at the Per-Flo Motel, her first betrayal of her marriage to Travis Word, she rides home with the windows down, the wind blowing. She exclaims, "'Glory Hallelujah!' I thought I had been born again" (225). For Grace, then, the conflict between physical and sacred pleasure results in a blurring of the line that distinguishes one from the other. The irony here is clear, as she has experienced so much anger and hurt at her father's "revival affairs," fueled by his similar attraction to the ecstatic. Yet in the midst of her own experience, Grace cannot so easily distinguish between sexual and spiritual ecstasy. Her moments of confusion on this point challenge the dichotomous view of human spirituality, the simple mandate that to find the sacred, one must deny the body.

I Have Been So Many People

Because of both her indoctrination and the pain her father's actions have caused her, Grace is troubled by having to make a choice between dutiful action and surrender to ecstasy. She considers herself hellbound because of her neglect of her marriage and daughters when she leaves Travis for Randy Newhouse, so it is clear that she does not fully trust physical pleasure as an avenue to heaven. Further, even spiritual ecstasy, in its emotional/physical climaxing, becomes suspect at times in her consideration of God's will for humanity—for example, Grace's recognition in Travis Word of the discipline her own father lacks. When Virgil leaves Grace in East Tennessee to take up with Carlean Combs, who has been "saved" at his revival, Grace finds great comfort in Travis's sense of duty. Travis is quite legalistic in his Christianity, and his condemnation of Virgil's actions offers her an escape from the hold her father's dogma has always had on her. "'I have been thinking about your father,' [Travis] announced in a preaching tone of voice, 'and as far as I am concerned, what he done passeth understanding'" (164). Travis's judgment is consistently supported by Biblical verse, as he has thoroughly studied God's Word:

> *I'll tell you what's the truth, from my own study of the Holy Word which I have read all my life, your daddy is flirting with fire, that's for sure. For it says real plain, "Do you suppose, O man, that when you judge those who do such things and yet do them yourself, you will escape the judgment of God?" And yet again, "For He will render to every man according to his works."* (164)

Unlike Virgil's pursuit of the ecstatic, Travis spends the majority of his time serving the community and his family, doing the necessary work of sustaining life: "'The Lord loves work,' Travis said. 'He loves a workingman'" (173). While Grace cannot sacrifice all earthly pleasure to such a concept of duty, she is greatly comforted by the image of God "as a rock, eternal and unchanging" (165). Her identity, as a result, is characterized by contradictory selves that somehow exist together: the devoted mother and faithful wife, the unquestioning believer and child of God, and the physical woman, finding joy in coconut cake, sex, the sun's warmth, and the feel of the wind.

Further complicating these apparently contradictory selves, Grace struggles mightily with the unresolved conflict between justice and grace. She has conceived of God, from the beginning, as a just God. This notion of God is a problem for her since, by her own claim, she is unworthy, unredeemable. She believes she is "lost," possibly for good. But she also feels resentful that the God, who she believes has judged her, has not judged her father, even after all the backsliding: "[H]e always comes back to God in the end, and apparently God is real glad to have him, every time. This is the part I can't figure out, how come God is so glad to have him. If I was God, I'd get tired of it" (164). Clearly it is Grace who has grown "tired of it"; thus, even though her own doom is part of the equation, she considers the appeal of a just God.

This concept of God does not totally satisfy Grace, though. Travis is so stern in his devotion to God that there is no room at all for flexibility. While he constantly serves his parishioners, he will not allow anyone to return his generosity. "'I pay my own way' was one of his mottoes" (207). Travis's concept of service is guided by a strict sense of accounting which at times overrides his value of community itself. And his inability to accept the legitimacy of his marital sexual relationship leaves Grace feeling that the relationship is cold and inorganic: "For a long time Travis's attitude toward bodily love did not seem too important, but then there came a time when it did, when I reckon my true nature came out too. For there are ways in which it is easier to live with a plaster saint like Daddy than with a real saint like Travis Word" (197). Thus she is not quite comfortable either with grace, as it refers both to the concept and to her own identity, or with the legalism of her religion. She lets go of neither, however, illuminating further the complex state of her identity.

Smith has, in mapping these complicated dualities and paradoxes, revealed much about the dilemmas of human spirituality, particularly in Protestant Christianity. However, her deconstruction of it has left many readers puzzled. How could these paradoxes be overcome in the act of Grace's returning to her old religion at the end of the novel, as some believe she has chosen to do? If, on the other hand, she has resolved her conflicted self by taking up a more feminine-based spirituality, as Byrd-Cook has suggested, how does this "mother"-based system answer the questions that Grace has grappled with all of her life? And finally, if she is not returning to her home or her Holiness religion for good, what *is* she going to do? Where is she going? Here, it is impossible to deny that the novel is flawed, that it leaves us in confusion with no clear purpose for doing so.

Yet, it is Smith's motif of fluidity that keeps the novel from falling apart. From the very beginning of the story, Grace's narrative is filled with expressions that characterize her experience as protean and unfixed, echoing this important quality of Ivy Rowe in *Fair and Tender Ladies*. Initiating the reader into her story, Grace recalls her mother's "pretty voice which always reminded me of running water, of Scrabble Creek falling down the mountain beside our house" (3). Scrabble Creek, in fact, becomes the center of the motif for the narrative as a whole. It provides a sensual memory of "home" for Grace, even when she has gone:

I Have Been So Many People

The mountain was steep beside our house, and the creek proceeded down it in a line of little waterfalls all the way to the road. Each waterfall had its own pool, some of them big enough to fish in, or to swim in when it got real hot...A great thrill would shoot through me whenever I held my nose and dunked myself all the way down under the cold rushing stream, even though Mama had said not to. (12)

Smith develops this powerful motif further by articulating the fluid qualities of wind. Just after her fourteenth birthday, in a moment of near ecstasy—associated with the natural, earthly world rather than the "other," transcendent world—Grace climbs to the top of Chimney Rock alone. "I took Daddy's jacket off, then took my shirt off too. I didn't know I was going to do it before it was done. The wind felt great on my chest and my back. I reached up and took off the rubber band that held my ponytail, and let the wind blow my hair all around" (58). The wind and her hair both flow, reflecting a moment of freedom for Grace. It is significant that while she is still atop Chimney Rock, she writes her name, "Gracie," in bold letters across the rock. Although she will search throughout the story to find her identity, she asserts it in this moment, a moment protean in quality, the fluid wind whipping around her.

Perhaps Grace's most profound articulation of the liquid state that characterizes her existence is expressed in her description of her early days of motherhood. For her, time seems to be suspended as she floats through the days with her daughters. "I felt like I had been there forever and ever, a thousand years. Life seemed to pass like a big slow river" (195). While Grace's sense of urgency and impulsiveness is reflected well in the "rushing waters" (272) of Scrabble Creek, her feeling of being taken through her life by a "big slow river" conveys with real impact the fluid context of her existence as a whole, and perhaps especially of her contradictory experiences with religion.

It is this protean quality, conveyed through Grace's narrative, that gives *Saving Grace* its edge. The reader's anxiety that the story will not hold together is maintained, yet not quite fulfilled, in the tension between the paradoxes and the fluidity. Here, Smith achieves what Roland Barthes describes as the "subversive edge" (7). This is not, in Smith's case, a political or religious subversion, but a narrative subversion. Barthes explains the wonderful effect of tension in such a work:

Two edges are created: an obedient, conformist, plagiarizing edge (the language is to be copied in its canonical state, as it had been established by schooling, good usage, literature,

> *culture), and another edge, mobile, blank [and I would add liquid] (ready to assume any contours)…Neither culture nor its destruction is erotic; it is the seam between them, the fault, the flaw, which become so.* (6-7)

Juxtaposed against Grace's seemingly contradictory reactions to experience, the narrative itself achieves an unfixed quality.

The novel is written in first person, from the point of view of Grace herself. She begins, "My name is Florida Grace Shepherd, Florida for the state I was born in, Grace for the grace of God" (3). She splits her story into five sections: Scrabble Creek, Traveling Light, I Settle Down, A New Paint Job, and The End of the Love Tour, each reflecting major phases in her life. Ostensibly, all is told in retrospect from a point in time after Grace's leaving Knoxville to return to Scrabble Creek. Upon her return, she runs into a childhood friend, Doyle Stacey, and in telling of this encounter, she reveals her temporal position in relation to the narrative's production: "That was a week or so ago, I forget exactly. I am losing track of the time up here. In fact, there is not any time up here now, not really, except for the day and the night, and the different light of the sun and the moon and the long white sweep of the snow" (267). Shifting here into present tense for the first time since the novel's introduction, Grace reveals that she has composed the narrative since her arrival at Scrabble Creek. This shift jostles the reader into recognition that Grace has told this story for a reason. She justifies the telling in the first pages of the narrative:

> *I never know where I'm going, and I never get there.*
>
> *I reckon I never did get there.*
>
> *This is why I have had to come back now, traveling these dusty old roads one more time. For I mean to tell my story, and I mean to tell the truth. I am a believer in the Word, and I am not going to flinch from telling it, not even the terrible things…I have got to find out who I am and what has happened to me, so that I can understand what is happening to me now, and what is going to happen to me next.* (4)

In a kind of spiritual hibernation at the Scrabble Creek house then, Grace sorts out her past by writing it, and in doing so, she is drawn to her mother's spirit and to Fannie's method for achieving spiritual ecstasy, handling live coals. Grace relates, "The Spirit comes down on me hard like a blow to the top of the head and runs all over my body like lightning" (271). Perhaps it is the coals that provide the vehicle for the ecstatic moment, or perhaps the writing of her story has brought Grace to it. At the least, one must acknowledge that her writing "the truth" has been a necessary part of her journey to understanding, as she expresses it, "who I am." The "truth" in this case is riddled with contradictions, and the narrative allows for them. Grace's use of the capitalized "Word" to describe her project reveals her insistence that her perhaps unorthodox story is as important to her identity as is God's Word.

By presenting the narration as such a powerful vehicle, Grace has merged the Word with ecstasy, the law with the transcendent moment. Thus, though we still do not receive a clear revelation of what *follows* this moment, whether Grace is going to testify to the world regarding her conversion or to end her physical life in an embrace of the spirit, *we* have been witness to her creation of a fluid narrative/identity capable of filling the breach of the dualistic flesh/spirit ideology. In spite of the discrepancy of Smith's conclusion to the novel, the reader is entranced by Grace's story, caught in the "seam" Smith has created between Grace's paradoxical position and the protean quality of her life as well as of the narrative itself.

Chapter Eight
Always the Storyteller's Story: *The Last Girls*

Lee Smith's *The Last Girls* is about multiple facets of human life: It is about the desire to overcome loneliness, the profound importance of friendship, the choice of either embracing risk and action or being responsible and safe, and the difficulty of facing mortality. But first and foremost, this novel is about the importance of stories to our notions of who we are. The narrative is characterized by multiple voices, multiple points of view: those of Harriet Holding and her college friends—Anna Todd, Courtney Gray, and Catherine Wilson; that of the late Baby Ballou, through her youthful raw poetry; that of Russell Hurt, Catherine's third husband; and that of Charlie Mahan, Baby's husband who has survived her. Yet, it is Harriet's perspective that most strongly shapes the novel. It is she who experiences the crucial transformation that, according to the girls' creative writing professor, Lucian Delgado, a good story must trace (353). And, most significantly, it is she who recognizes that the traditional narrative structures, "beginning, middle, and end; conflict, complication, and resolution... don't fit Harriet's life, or the lives of any women she knows" (20-21). Harriet's recognition of this problem leads her to teach creative writing, encouraging struggling women to tell their stories in ways that work for them, and to embrace the notion that each story belongs to the storyteller, one's story shaping one's identity in powerful ways.

The plot of this novel is, to some extent, a playful one: Harriet and eleven of her college friends decided, during their years at Mary Scott College in Virginia, to "do the Huck Finn trip" down the Mississippi River on a raft, attempting to relive Twain's story, which had made a strong impression on them, as it has on American identity at large. Now, roughly thirty-five years later, four of them have reunited to take the trip again, this time with Baby Ballou's ashes, at the request of her husband Charlie Mahan. This second trip is different from the first in many ways: This time it is a cruise on a luxury riverboat, the Belle of Natchez, rather than the crudely built raft, the Daisy Pickett; and this time, though Russell deems them "the last girls" (71), toasting them on the river boat, the women are substantially different people than they were as young, naïve coeds. Since those early days, life has taken its toll on each of them. However, each still wrestles, on some level, with the issues of identity she negotiated back then. And these women are not so jaded and tough now that this second trip occurs without its own profound effects on them. If the first trip was an exposure to their Southern history, to Southern land and waterscapes, and to their own roles in the complicated world around them, this second trip provides an opportunity to reimagine themselves, their desires, and their roles in what is, in many ways, a world different from the one they explored so many years ago. This time, rather than being preoccupied with becoming a "great writer," each character regards and shapes her own "story," her self-narrative, and its meanings, recognizing the power of these narratives to trap or transform.

In addition to contrasting the original Huck Finn trip with the later one, revealing the evolution of each character over time, Smith unfolds the novel's details by way of a third-person limited omniscient narrative which shifts, from section to section, to reflect the individual consciousnesses of these characters. By way of this strategy, Smith conveys the personal story of each character with implied authenticity, these "little narratives" resisting appropriation by the "grand narratives" of Southern culture or contemporary society. Smith's decision to employ multiple points of view in this way supports the novel's assertion that each story belongs to the teller and that in sustaining such a model, contemporary society can both preserve story as a crucial factor of culture and resist imposing narrative traps on its subjects.

The "Huck Finn trip" itself provides a pertinent setting for exploring this point. The twelve girls on the original trip were all students in Mary Scott College's writing program. Many of them had serious aspirations of being writers, and when Baby suggested during their Great Authors class that it would be great fun to go down the Mississippi River on a raft like Huck Finn, these girls caught the fever. Unsurprisingly, their experience turned out to be quite different from Huck's: undergoing their own "hero's journey" with the knowledge that "[i]f anything really bad happened to them…they could call up somebody's parents collect, and the parents would come and fix things," they became regional celebrities but did not find the opportunities for philosophical and moral growth that Huck encounters (18). Yet their celebrity is only one reason for their failure to repeat Huck's trip. As Ivy Rowe (*Fair and Tender Ladies*) understands, the authentic "Huck Finn experience" is not possible for a girl. Ivy explains, "I had half a mind to try and go as a boy and ride a raft myself, but I said, Now Ivy you know you can not, you will never get away with it" (86). She is adolescent, and is aware of the men staring at her body "threw my dress" (*Fair and Tender Ladies* 80). The experience of Ivy and of the Mary Scott girls could not be divorced from their sexualized bodies, as Huck's seems to be. In spite of the attention Harriet and her friends drew on that first trip and the occasional escapade with local boys they encountered, including Civil Rights workers in Natchez, Mississippi, several of them did manage to spend time writing: Harriet, a coming-of-age novel that she never finished; Anna Todd, a novel that preceded her eventual grand success as a romance writer; and Baby, her scribbled poetry. Yet, their adventure certainly did not follow the narrative pattern of the "great American novel" that had inspired it.

Embarking on this second trip, the Mary Scott "girls" no longer harbor naïve dreams of playing Huck or of joining the ranks of the "great authors." Yet, the travelers still recognize the power of stories, whether in the form of narrative, photography, or sculpture. And this trip drives home for them the importance of retaining ownership of one's story in order to understand and obtain agency over one's identity. As middle-aged women whose lives have become increasingly complicated, they illustrate this importance by their various individual "tales."

Anna Todd, née Annie Stokes, has perfected the formula romance novel to achieve both wealth and fame. While it may seem she has sold out her dream of serious poetry and fiction-writing, it has not been quite so simple. Perhaps the best writer of all the girls during college, Anna wrote then about the harsh life of her West Virginia childhood. Her first poem for Mr. Holland's Introduction to Poetry class was "Little Finger Bones," about a mountain man who kills his unfaithful wife and uses her "little finger bones" as banjo frets (120). Anna's professors and peers were fascinated by her dark and provocative subjects, and she eventually even won a fellowship for her creative senior thesis, "the first four chapters of the novel she was writing" (139).

Yet the events that followed her successful undergraduate years brought a heavy weight to her existence. She delivered a stillborn child, and when a literary agent responded with enthusiasm to a novel she had written, her husband Kenneth could not deal with the threat to his ego and left her for another PhD graduate student. Her novel was ultimately rejected by publishers after all, because it was "too raw" (145). When she finally sought solace from her fellow employees in the filing unit at the hospital, her friend Cindy offered her "a paperback named *Mortal Passions*, featuring a graveyard with a castle in the background" (147). Cindy advised Anna, "I'm not kidding, just read it, you'll see, it'll make you feel *so much better...*" (147).

A talented writer, Anna very quickly sees that not only can she achieve great success with the romance formula, but she can also reside, through her writing, in a world that meets both the human desire for order and the yearning for sensation, an escape from the banality and angst of ordinary human existence. Transformed by her experiences, Anna now insists that "[a] story must have a plot" (79). In a conversation on the riverboat, as Harriet describes the life stories her COMEBACK! students are writing, Anna chides, "Who would want to read about people like that?" (78). For Anna, writing is a way to remake the world, to take authorial control over human experience and adopt a romantic view of existence. Although this impulse to control narrative does help her cope, the grand narrative of romance distances her from the experiences that have formed her and obscures her own personal story.

Courtney Gray constructs a similarly artificial coping mechanism for herself through her carefully designed scrapbooks, which feature many of her own photographs. Always one to respect social rules, Courtney has thrived in situations where she could organize and manage life's details. Her scrapbooks, then, are a natural manifestation of that managerial personality, also serving as an outlet for her creativity and her strong desire to capture her memories. Unlike Katie Cocker's intense and authentic memories in *The Devil's Dream*, Courtney's are carefully selected and crafted to reflect the life Courtney imagines hers should be. Through the scrapbooks, she constructs a visual narrative of her life, casting her preferred meanings on it in the process.

Notably, Courtney's life has not turned out the way she planned. Having grown up poor, she imagined her marriage to Henry "Hawk" Ralston as ideal. She has done everything "right," giving birth to two sons and a daughter, becoming the perfect hostess, volunteering in her Raleigh, North Carolina, community at the library, the rape crisis center, and her church. Along the way, she has given up her hopes of painting, making her home her canvas instead. In spite of her commitment to her family and the standards of upper-crust Southern society, however, Hawk has kept other women since early in their marriage, her daughter is now traveling with a rock band, and her son Jeremy is "troubled," having dropped out of college and moved to Boulder, Colorado, to work in a bookstore (48). Courtney's decision to have an affair with the zany florist and designer, Gene Minor, is completely out of character for her, she feels, but this unconventional relationship helps her break out of the strict role into which she and society's expectations (the "grand narrative") have trapped her. She and the three hundred pound Gene follow impulses like closing down Gene's flower shop and spending the day in bed or eating enchiladas and drinking margaritas in the nude. "If it's not fun, don't do it!"—this is Gene's motto (46). Of course, Gene does not appear in Courtney's scrapbooks. The discrepancy between her idealized "story" and her fling with Gene is too great for her to reconcile, though her time with Gene has helped her avoid the breakdown toward which her polished performance was leading her. Yet perhaps even more important to her psychological survival than her affair is the photography she has done all along. She does keep her photographs in her scrapbooks, strategically organized to represent the various aspects of her idealized life: her children in their soccer uniforms and tutus, Hawk with a fish he caught once off the coast of Cozumel (47-8). But in her photography itself, she merges her desire for order with her creative drive. Her visit to a St. Francisville cemetery, during this most recent riverboat trip, reveals her in her element:

> *The old live oaks stretch out their long furry gray arms to form a canopy over an ancient little brick Episcopal church and all the old graves that disappear into the shadows there at the edge of the frame: click. White stones rising into consciousness like ideas, like memories, like ghosts; darker, older stones with names too faint to read, souls long lost to time. Click. Time has stopped dead in here, this high dim leafy tent where it's always cool with a little breeze that makes a sound you can almost hear as it sighs through the Spanish moss. The trees are so tall that they creak, leaning toward each other, telling old, old secrets.* (284)

Her recent decision to return to Hawk, now suffering from what appears to be the onset of Alzheimer's, rather than leaving him for Gene Minor, seems apt, given that Hawk is part of her "ideal life," the one she has logged in her scrapbooks. Gene has been an escape, a fantasy. He has helped her achieve perspective, but the affair will not provide her, ultimately, with a sustainable way to understand and sustain herself. She achieves momentary clarity of perspective as she walks through the cemetery. As she takes respite in the old Episcopal church next to the cemetery, she realizes that in this life that has not necessarily followed her plan, she has still glimpsed the sacred: "And yet, as now, she has always been thinking of something else—who to invite to dinner, what to wear…the little things of life that are holy, too, or so she has always thought" (287). Courtney's scrapbooks chronicle the story of her life as she imagines it, generally idealized; yet in her photographs, she shapes a parallel and more authentic story. Here, she unabashedly takes authorial control and reflects a self that is, on some level, independent of the "grand narrative."

Catherine Wilson, too, has chosen an outlet other than writing for her story. For her, sculpture has been a way to find and own the narrative of her *self*. Unlike Courtney, however, Catherine has always been a free spirit whose creativity ruled her daily life. Even her commitment to those early creative writing classes was easily overridden by projects she was doing in the theatre or in sculpture classes. She often came to class dusty, late, and distracted. Catherine's experience with life has been based in images rather than words:

> *The way they'll come to her at the most surprising times and places in the midst of life, when she's doing something else completely, such as unloading the washing machine, and suddenly she'll see a shape in her mind's eye, a triangle, for instance, then a chair made out of triangles, and she'll just have to leave the wash or whatever she's doing, and go out to the shop and start making that chair. All her life, Catherine has been easily overtaken: by her husbands, by her children, by images and ideas, by life itself.* (188)

Although this tendency to let herself be "overtaken" has often led to heartbreak, it has ultimately been the source of her creativity, her salvation.

As a young woman, still under the assumption that her life would look like the models offered her, Catherine planned her first wedding in an attempt to fulfill her Birmingham family's expectations, but it was not long into her marriage with Howie that he became tense and annoyed with her. Unable to live out this "story," Catherine took their son William and left Howie for an emergency room intern, Steve Rosenthal, with whom she lived for only a few years, until he was killed during a convenience store robbery. She had had three children with Steve, so she worked various jobs to raise them, along with her oldest son, William. Even under such heavy economic pressures, though, Catherine's strong creative drive pushed to the surface: "Eventually she started making these big concrete women with mosaic tile dresses and hats. People bought them for crazy prices" (198). The images in her head have always dominated the external narratives that might otherwise have overshadowed them.

Catherine has reflected her life in images associated with her experience of it; her sculptures have become her story. It is not the neat narrative of Anna's novels or of Courtney's scrapbooks, certainly. It is not even collected, but is scattered—in the form of moon-shaped stepping stones, fountains, tables, and statues—all over the South. Her old creative writing teacher, Mr. Gaines, would perhaps not even have recognized these images as a story. Yet, it seems that Harriet, now a creative writing teacher herself, would.

On the Belle of Natchez, in a discussion with Anna about stories, Harriet becomes heated in her insistence that not all stories can be told in the old conventional formulas. At a Virginia community college, Harriet teaches a "Write for Your Life" workshop, where she "tries to help her students tell their own life stories[.] [S]he has learned that there are more ways to tell stories than she could ever have dreamed. And all the stories are different" (21). Though Mr. Gaines presented Huck Finn to the girls as "an American Odysseus off on an archetypal journey, the oldest plot of all" (21), what Harriet connected with in this story was Huck's loneliness (13). Harriet has always tended to get "all wrought up" (13), as she does when Anna insists that it is not good for Harriet's students to "write their own life stories" (78). "Well, I've been working with these women for years, and I disagree." Harriet won't let it go. "I mean, these may not be stories in the way you think of a story, I realize, with a strong plot and all, of course" (78). Underneath Harriet's discussion of the conventions of narrative is her absolute *need* to preserve the possibility of nontraditional, unappropriated story. Now that she is formulating her own story, she recognizes that it cannot be told in the formulas offered by her creative writing teachers or the "great authors," and she recognizes that telling this story is part of her own journey to understanding and determining her self.

Even Anna seems to acknowledge Harriet's point about life's non-linear quality later, when she stands on the deck with Harriet, who remarks on how huge the river is, marveling that they had the courage, as young girls, to embark on such a dangerous journey. Harriet says, "I guess you can't tell how big things are when you're right in the middle of them anyway." Anna answers, "Thank God! . . . It's just as well. If we could ever really see what we're doing, then we'd never do any of it, I imagine" (81). The details of Harriet's background suggest that perhaps this is what has kept her from acting on her impulses all her life. She *could* see how big it was. While others took risks and she stood by as an observer, her fear of "diving in" kept her from claiming her part as protagonist of her own story. She absorbed the enormity of life as it occurred, thus she could only bear to live it vicariously, through friends like Baby Ballou. Now, in her capacity as creative writing teacher, this perspective gives her an advantage. This role, taken on in her middle age, reflects her dawning recognition of story's importance to a person's identity and, next, leads her to the turning point at which she can claim and write her own story *and* identity.

I Have Been So Many People

    For someone who has experienced life second-hand, Harriet has observed a great deal. Harriet has taken care of people all her life: She took care of her sickly sister, Jill; she took care of Baby, and then Baby and Jefferson Carr as a couple, though she was in love with Jefferson herself. Baby's "response to everything" was "YES!" (15), which gives Harriet a window onto life through someone she can see is on a destructive path but who embraces risk and excitement. A "love child" (31) herself, Harriet takes consequences too seriously to risk hurting anyone. Yet, she feels she is the one who has to carry the burden of all that has happened to her and to those she loves. "Only Harriet seemed doomed to remember and remember and remember, to remember everything" (132). Perhaps it is as complements then that she and Baby are drawn to each other. Baby tells Harriet she is her best friend and insists that Harriet come with her on many of her escapades. It is to Harriet that Charlie sends Baby's ashes and his letter about Baby's last days. In this missive, he acknowledges that "times of trouble" (358) still came periodically with Baby, but he says that on her newest medication, Baby had been feeling, in the days before her accident, "the best she had felt in years" (359). His words are a gift to the travelers, though Harriet is never sure whether or not Baby's death was an accident, as Charlie represents it. For Harriet knows, by this point, that "it's always the storyteller's story" (363). It is Harriet who rescued Baby's poems from a hotel room table during their first voyage down the Mississippi; it is she who treasures them as a legacy of the girl whose terrible insights into life haunted her all of her relatively short life.

    These poems, which lace the narrative, reveal Baby's need to tell her own story, even if it was to then end it in possible suicide. The writing seems always to have been therapeutic for her, a way to unleash her violent and torturous thoughts, her knowledge of a family history that would not be welcomed by most readers. These poems record images of her scandalous mother, who "wore red short shorts / and high-heeled sandals" ("MAMA I" 178) and "drank gin like water / all day long / in the pink glass goblet / with the twisted stem" ("MAMA II" 179); of her older brother Ricky, who committed suicide at a young age; and of Baby's time in a mental hospital, where "They all think / They're Jesus" ("DAY ROOM" 180). Baby reveals, in "PLEA," that the pedestal on which the devoted and pure-hearted Jefferson has placed her is "hurting [her] ass" (179). Interestingly, Baby seems to have been, on the first trip, provoked by the same St. Francisville cemetery that captures Courtney's imagination on the second. But, in "AT THE CEMETERY," Baby focused on "the tree cut down / in the prime of life / draped with a shroud" (288), ostensibly referring to her longed-for brother, Ricky.

Yet even with her scribblings, Baby's story becomes appropriated by others. Each of her friends on the riverboat trip interprets her: Anna sees Baby as always her competitor, as a writer and as Mr. Gaines's lover; Courtney is easily convinced by Charlie's letter that Baby was happy until her "accident" occurred (363); Catherine reacts with an uncertainty that allows her to entertain the most positive possibility: "Well, you never know...I mean who ever knows what anybody's marriage is really like? Or what's going on in anybody's head? But it certainly doesn't sound like she killed herself. I guess it really was an accident, don't you think so?" (363) Harriet, who has carried around her knowledge of Baby all these years, suspects it is more complicated than this. She understands that while "readers" will interpret one's story, the story itself belongs to the teller.

In an interesting twist, the event of Baby's death has become an event of each of her friends' own stories. At Baby's memorial service, Anna weeps, but not for Baby—for Lou, the man who had truly loved her even after she had become jaded, until his death of a heart attack. She thinks, "It's true that when anyone dies the other dead rise up and die all over again" (367). These specters live for Anna in spite of her success in reiterating over and over again the romantic formula in her fiction. Similarly, threatening Courtney's positive spin on Baby's late life and death, she feels "this terrible sense of desolation sweeping over her suddenly" (366). Then she pulls out her camera to photograph this moment, thinking about how she'll return to care for Hawk in his time of need. Catherine, who has been faced with her own mortality at the discovery of a lump in her breast during the riverboat trip, turns to Russell, from whom she has felt so alienated recently, and takes comfort in his presence, in spite of the idiosyncrasies she has found so annoying lately. Each of these characters is still crafting her story, still making decisions at each turn about whether to link up with the "grand narrative" or to choose a more honest plot turn. In Smith's portrayals of these authorial choices, she suggests that the more independent the story is of the "grand narrative" the more power it has to liberate and sustain the teller.

At the memorial service, Harriet is most affected of all, it seems. As a result of Baby's death, Harriet's personal narrative, as she has constructed it for herself, has shifted, disorienting her. She has felt, since she first heard of Baby's death, that it was suicide, that if only Harriet had helped Baby and Jefferson get back together after their last breakup, instead of sleeping with him—the one truly passionate act of Harriet's life—Baby would still be alive. But now she is not sure: "[S]he wonders how she could ever have been so egotistical as to presume, even for one minute, that her actions were of such importance...Baby grew up, that's all—while she, Harriet, did not" (369). Harriet understands, too, that she will never know for sure what really happened to Baby, but somehow this rupture in her narrative has liberated her to pursue the authentic possibilities. Pete, the Riverlorian, puts his arm around her, an unspoken question of whether she will join him for a weekend in New Orleans, here at the end of the journey. "'Okay,' she says. 'Okay'" (370). This answer shows Harriet reconciling herself to the YES she loved so much in Baby, to the chance she still has to live life instead of only observing it. Her evolving relationship to writing, along with her newfound courage to participate as protagonist of her own story, leaves Harriet poised to become agent of her own narrative and her own identity.

The Last Girls covers a great deal of territory, from the 1950s to the 1990s, from the Mississippi to the east coast, and from childhood love to marital complexity. Yet, what generates the work's impact is its adept exploration of story and how it shapes human experience. The key figure of the novel, Harriet understands that a story depends upon its teller, that even though the story may never capture an empirically provable reality, it does provide the teller with a level of control over her/his identity and history. She recognizes that each teller's rendition of an experience will be different, and her acceptance of these incongruities casts light on the novel as a whole, a collage of narratives from various perspectives, which together offer us a rich portrayal of these "last girls."

Chapter Nine
"We are all just passing through": Contingency in *On Agate Hill*

"My family is a dead family, and this is not my home, for I am a refugee girl" (7). So claims thirteen-year-old Molly Petree as she begins to record her story in the new diary given her by the preacher's wife, Nora Gwyn. From the narrative's beginning, both of Molly's parents are deceased: her father, a Confederate officer, was killed at Bentonville, and her mother, former "belle" of Perdido plantation in South Carolina, died from an illness she contracted after taking refuge from the war with her younger children at Agate Hill, home of her first cousin, "Uncle" Junius. Molly is, indeed, as she refers to herself, a refugee and an orphan. Yet, this status, taken up almost purposefully by Molly, represents more than her immediate condition. As an orphan and a refugee, Molly subverts the assumptions of who she ought to be, given her family and class. Although this position leaves her lonely and sometimes vulnerable, Molly insists on engaging in the world more directly than her Southern aristocratic identity would allow. In this choice, she rejects even the social and economic privilege assumed in her "noble" identity: "I want to feel everything Dear Diary. I want to feel everything there is. I do not want to be a lady" (73). As Molly comes to understand that reality can only be authentically and meaningfully experienced as contingent, rather than as fixed or absolute, Smith illuminates a central contemporary dilemma, the choice between preserving the myths of cultural ideals and taking one's chances without them. Ultimately, Smith recommends the latter.

Through this novel, Smith offers a perspective which may help us negotiate our rich and often troubled pasts in order to embrace the present and our contingent selves. Like most people, Molly craves stability and community, but she refuses to accept them at the cost of psychological entrapment. Instead she approaches life through an awareness of its contingent quality, privileging experience over ideals. Smith develops this point further through her use of a multiple-sourced narrative: She employs a variety of narrative perspectives, through diaries, letters, school and legal documents, even objects. The story's frame, provided by Tuscany Miller's twentieth century documentary studies project on the box of Molly's "phenomena" which she found in Molly's Agate Hill cubby hole, emphasizes not only that the *meaning* of our realities is inherently contingent, but also that to grow, we must return to interpret those realities again and again.

Tracing Molly's life, from her arrival at Agate Hill at thirteen years of age to her impending death as a 68-year-old woman at this family homeplace, the novel relates her evolution from a child, struggling with society's imposed hierarchies and values—represented to a great extent by Agate Hill itself—to a mature and liberated aging woman, once again living at Agate Hill but no longer dominated by it. What has changed for Molly? What has lifted from her shoulders the weight of Southern ideology? The key is her recognition that all is contingent. As Molly develops over the course of the novel, so does her understanding that meaning—of her family's status, of the Agate Hill estate, of her mysterious benefactor Simon Black, and of her own identity—shifts according to context. It is her acknowledgement of Simon Black as a human being—rather than a force that threatens her independence—that signals her final moment of illumination, the moment when she sees fully that home, family, and self are not fixed entities, captured by abstract ideals, but are, instead, contingent factors only to be meaningfully understood by experience. Further, Tuscany's exploration of Molly's diary—at an Agate Hill newly renovated into a successful bed and breakfast, complete with "fancy decorator cows" (362) in the fields—leads to her own growth: "[M]y horizons have been expanded by the contents of this box" (362). Initially terrified by her father's sex change and the complete redefinition of her family and self as a result, Tuscany has, after immersing herself in (and constructing) Molly's "story," learned to love her father as an evolving person and to accept their relationship as a rich and necessarily contingent aspect of her life.

Molly's early claim in her diary that "this is not my home" (7) reveals her sense of displacement and indicates very early in the novel that the story is a decentered one. By decentered, I refer once again to Jacques Derrida's notion of this term, which Stanley Trachtenberg discusses in his introduction to *Critical Essays on American Postmodernism*. Trachtenberg notes that, responding to the Platonic notion of an ideal center or origin, Derrida argues meaning is generated in a more dispersed way than that which posits meaning as ever-referent to the original, or the central (10). Elaborating on Derrida's theory, Denis Donoghue explains,

> *[T]he entire [Western] philosophic tradition is infatuated with the notion of a first moment, authentic and paternal...so the best answer to our infatuation with voices and presences is writing—writing stripped of all delusions. Writing in this sense has given up yearning for a lost father, it knows that it is an orphan.* (33-34)

Smith's portrayal of Molly as a literal orphan is no accident. Even thirteen-year-old Molly understands that her mother represents this infatuation with "origin"—Alice's being traced to her family's plantation, Perdido. On her deathbed, as Molly watched over her, Alice related Molly's history to her "in that awful whisper which went on and on through the long hot nights" (7-8). Molly admits only in her diary, "I loved mamma. But I was GLAD when she died" (8). Orphaned, Molly is, she perceives, free of that whispered history. Trachtenberg and Donoghue acknowledge in their discussions both the anxieties and liberation inherent in Derrida's vision of a decentered reality. Smith's narrative illustrates this dual effect of the postmodern condition, specifically in the context of a traditional Southern culture. Molly suffers often as "an orphan girl, loose in the world" (198), with no consistent supervision or friendship, yet because of this experience, she comes to see that there *is* no center to decipher, that the self is formed and maintained under conditions more complex than this, and thus there is no completely stable origin, not cultural, personal, regional, or familial. While one may feel meaningfully connected to a place, an event, or a person, Smith notes, the traditional Southern notions of home, family, and self are romantic and imagined, artificially protecting false and often irrelevant notions of identity.

Of course, critics of this postmodern insistence on reality's contingency allege that it leaves us with no basis for ethical action or for community. This complaint is a common one among Southern preservationists. How can we hope to "do right," connect meaningfully in community, or even remember who we are, without a common history, a common set of ideals, or a stable sense of what is noble and honorable? Molly's Aunt Cecelia is certainly committed to serving as a bastion of traditional Southern values; according to Molly, she "specializes in rising to the occasion and keeping up standards" (46). Arriving at Agate Hill from Alabama with her granddaughter and charge Mary White, Cecelia takes it upon herself to save the estate from its impending dissolution, as its integrity is threatened not only by post-war economic changes but also by the widowed tenant worker Selena Vogel, who intends to marry Cecelia's dying brother Junius. Cecelia is distraught over what she judges to be her brother's severe lapse in judgment.

Yet John Rothfork, in a defense of Richard Rorty's expression of postmodern ethics, argues that postmodern assumption of contingency is not without a strong commitment to both ethical behavior and community. Rothfork explains,

> *Those of us who trust Rorty's advice do not expect people involved in such experiments to be unprincipled nor to be mired in the philosophic swamp of moral relativism. We expect them to discover the principles that are important in their lives through their own experience rather than by taking principles off the shelf; out of some philosophy text or from a sermon or political speech...Idealists, or those who are devoted to transcendentals, discount the value of personal knowledge or knowledge as performance—operational knowledge acquired in actual communities and in lowly popular culture, such as television—because it threatens to vitiate metaphysical principles and, worse yet, to dissolve principles in contingency and ambiguity.* (n.p.)

In the face of these accusations, Rothfork counters,

> *Pragmatism does not concede to these terms. The pragmatist lives in multiple, concrete communities. Rorty says that in losing faith in the cosmic structure, it "does not seem to us to entail that we face an abyss, but merely that we face a range of choices" about which actual communities to become involved in (Papers, 2: 132). We say, it is the fundamentalist who cannot discern paradigm boundaries and consequently insists that there are none; that it is all or nothing.* (n.p.)

David J. Gordon clarifies the insecurities that lead to the sort of fundamentalism Rothfork refers to. In his discussion of Iris Murdoch's fiction and her assertion that contingency must be accepted for the sake of true morality, Gordon presents the traditional notion of *history* as an example of illusion imagined for the sake of "consolation" (116), "[imposing] pattern upon something which might otherwise seem intolerably chancy or incomplete" (Murdoch as qtd. by Gordon 117). Lindsey Tucker elaborates on humanity's difficulty negotiating a world characterized much more by chance than by pattern: "It [human experience] is rich, complicated, external, and—above all—contingent. Indeed, it is its contingent nature, its complexity, particularity, and messiness that the fantasy-ridden egotists fear and attempt to control by imposing form and pattern upon it" (4). Gordon and Tucker focus, like

Rorty and Rothfork, on the damage resulting from those false patterns that hinder real understanding and, for Murdoch, authentic goodness.

Unlike Murdoch, Smith's interest is not so much morality as meaningful living. As Molly implies, idealism is a buffer or mediator between us and experience/reality. To really *feel*, she must drop the shield of her position as "lady." This new, vulnerable state brings her into the moment, into closer contact with reality, which shifts constantly according to context. In this mode of negotiating reality and values, Smith does not give up the notions of home, family, and self; rather she frames them as factors contingent on conditions of perception, including time, space, and social environment. By portraying in shifting contexts Molly, Agate Hill, Simon Black and—in the narrative frame—Tuscany Miller's family, Smith illuminates the point that while places, people, memories and a sense of self are crucial, to experience them most authentically and thus to grow, we must be willing to give up the notion of them as permanent and to, instead, commit to a process of constant re-examination and reinterpretation.

As an adolescent, Molly is already dissatisfied with the meanings that have been imposed on her. From the story's beginning, she chafes at the restrictions she perceives to be part of her role as a Southern "lady." Her mother, Alice Heart Petree, strove until her untimely death to inculcate Molly with the codes of aristocratic womanhood: "Horses sweat, men perspire, ladies glow" (30), she told her daughter. Yet, as Molly notes, "that was back before ladies worked" (30). After her mother's death, as the Agate Hill estate diminishes in financial value, even Molly must pitch in and work to help sustain its minimal function. She recognizes, even as a girl, that as one's station changes, one's identity is revised as well. When Selena Vogel and her children move into the main house, Selena takes up the position as "lady" of the house, though her background as a laborer drives her to do much of the physical work herself to realize her vision for the old house and for her own status. Molly resents the imposition, but she is not as disturbed as many members of the community, and even the estate's black servants, at Selena's breach of class distinction. For example, Molly refuses to wear the camisole her mother left behind because she understands that with the position of "lady" come burdens of behavioral restriction she is not willing to accept. Instead, she relinquishes the camisole to Selena's daughter, Victoria. Old Bess, once a slave for Alice's family and still a loyal servant, tries to persuade Molly to keep it, since Junius's deceased wife Fannie "done save these things for you by the hardest." Molly replies, "I don't care" (31). At this, "Old Bess turned to look at me hard," Molly notes (31). Smith implies here that for most people, unlike Molly, the shifting social structure is perceived as a threat to self and to moral codes. Selena's "rise to power," tenuous as it is, is so frightening to the community (and understandably so for those beneath her in the social hierarchy) that after Junius dies, Liddy and Washington (Molly's beloved childhood playmate) leave Agate Hill (112). Yet, as devastating as this departure is for Molly, she does not fall back on her family's status as a way to regain stability. Though she could easily take up with Aunt Cecelia as a path back into relative gentility, she rejects this possibility in lieu of *feeling*. Illustrating this rejection, when Molly and her cousin Mary White ride in the carriage with Aunt Cecelia to Hillsborough to see friends and to attend the tableaux vivants, Aunt Cecelia tells her to put her bonnet back on or she will get freckles. "<u>I dont care</u>, I called back...I did not even bother to say, I dont care if I get freckles or not because I am not going to be a lady, I would rather die than be a lady like you" (66-7). Molly understands, at least subconsciously, that her family's position as *center* or *origin* is a heavy weight to bear. Further, it is inherently false, contingent on many factors, not the least of which is *money* in an unstable economy.

Beyond family position, Agate Hill, so full of meaning for its inhabitants and neighbors, is also suspect to Molly as a center. As much as Molly resists leaving her "ghosts" (136) when Simon Black and Agnes Rutherford come to take her to the Gatewood Academy boarding school, Molly has always seen through the façade of Agate Hill as a symbol of what is good and noble in the South. Taking refuge under the mahogany dining room table one day, after Selena has disciplined her by shaking her "until my teeth rattled in my head" (38), Molly notices the table foot next to her, "a huge mahogany claw that had seized a mahogany ball. Its talons were big and sharp" (38). She jumps up to leave, suddenly desperate to get outside, "But first I looked back down at the table, thinking, <u>It is like this house, it looks so fancy and fine, but it is all ugly underneath, it is that mean cruel claw</u>" (38).

Similarly, Molly's impression of the ever-evolving Agate Hill includes Mary White's and her encounter with the lynched and burned man when they visit Four Oaks to take Mama Marie some custard. Delighted to be traipsing in the woods and fields of the estate and entranced much of the time by their own games and imaginings, the two young cousins fail to see the hanging corpse on the way there, but on the way back, they are stunned as they come upon it, "a large negro, very black, with his swollen head drooped over to one side and his mouth open and his tongue out, eyes naught but bloody holes" (79). As Molly and Mary White discuss the terrible sight, Molly informs her cousin that "It was the Klu Klux. You know it was" (80). Molly, who has read Uncle Junius's newspapers, knows of other incidents like this one, and she unintentionally upsets the fragile Mary White further by telling her of them. But when they return home and tell Aunt Cecelia they have seen the lynched man, "Aunt Cecelia shook Mary White like a rag doll. <u>You did not</u>, she said. Listen to me. Girls you did not see that, do you hear me? I wont hear another word about it. Not another word. Do you understand?" (81) Molly, however, will not concede to practice this kind of history-making: "<u>Yes</u>, I said, but I dont, and I am going to tell Uncle Junius anyway" (81). Ostensibly, Molly is not gratified with any action by the magistrate, as she has hoped, for no more is reported on the subject, but she preserves the incident in her diary and remembers it always, for as she notes, "I remember everything" (194).

Agate Hill is undoubtedly "home" in this novel. Molly begins her story here, and after lighting "like the ruby-throated hummingbird" (7) as a student at Gatewood Academy, then as a teacher at the Bobcat School in Grassy Creek, North Carolina, and later as Jacky Jarvis's wife up on Plain View, North Carolina, she returns to Agate Hill. Yet, it would not do to state that this is a return to her "origins." Most of the outbuildings and main house are now crumbling, the main house barely habitable. None of the people she knew resides there now either. Instead, Simon Black, who purchased the house for Selena and kept it when she left, is dying there, cared for by his Brazilian servant and devotee, Henry, and the malformed child of Selena, Junie, who, now grown, inhabits the place like a sweet but feral animal. Almost surreal in this state, Agate Hill cannot function as a center for Molly, and to further emphasize its decentered position, after Simon dies, Molly, Junie and Henry move out of the main house and into the tenant house and Liddy's kitchen house, near the garden where they spend much of their time. In this move, the three give the main house, "Agate Hill[,] over to its ghosts for good" (350).

In Smith's hands, the rendering of Agate Hill, both the plantation where Molly's family found refuge and the decaying estate to which she returns, is of a place contingent and evolving, with a history defined not in the linear mythological terms of a traditional South but by its material conditions at particular points in time. When Molly, Henry, and Junie attempt to bury Simon Black in the family cemetery at Four Oaks, they are turned away, as that part of the original estate is now a privately owned country club and golf course. They bring the body back "home" then, resigned to burying Simon in the "garden plot beside Liddy's kitchen house" (350). Yet, here, too, they are met with material obstacles: the bones of the tenant farmer, Mr. Vogel, killed by Selena long ago—Molly understands now—so Selena could pursue Uncle Junius and her ambitions. These long-buried bones reveal Smith's landscape as one consistent with that of other important Southern women writers discussed by Patricia Yaeger in *Dirt and Desire*, those who reject the "white-washed" images of the South to portray instead "landscapes loaded with trauma unspoken, with bodies unhealed or uncared for, with racial melancholia" (18). Harboring Mr. Vogel's bones, the lynched, burned black man, and the "Yankee" hand eventually preserved in Molly's and Mary White's box of phenomena, Smith's landscape literally embodies the material history of Agate Hill. Accepting the contingent circumstance of Simon's burial, Molly, Juney, and Henry finally choose a place "by the miller's stone" (350). There is no marker for the grave, only sunflowers that "come back bigger every year" (350). By now, contingency is such a central aspect of Molly's perspective that the lack of a grave stone gives her no sense of loss.

We are not left with this decaying image of Agate Hill, of course, but rather the final picture is Tuscany's, of the renovated and commercialized version of this estate, a bed and breakfast featuring rooms named after the people described in Molly's papers: Uncle Junius's Study, Aunt Fannie's Sewing Room, and Molly Petree's Cubbyhole. As in *Oral History*, with the Cantrell land and homeplace turned ultimately into an amusement park, we are forced to acknowledge that context is crucial to really experiencing a place at any given moment in time. Alice Heart Petree and Aunt Cecelia would surely turn over in their graves to see Agate Hill rendered so. Yet, for Tuscany this lesson is crucial. She can now consider her father and her evolving family through this same lens of contingency and can finally recognize the love and intimacy possible beyond her previously rigid ideals of family and identity.

Unlike Tuscany, who first appears to us as a somewhat closed-minded young college drop-out, angry at her father for shaking her ideals, Molly has from the beginning been aware, on some level, of the contingent nature of reality. Yet even Molly's ability to treat people as subject to contingency is imperfect. This point is revealed by her resistance to her "benefactor," Simon Black. In fact, Black serves in the narrative as a symbol of mystery, misinterpreted by Molly and others as a result of their defaulting to the abstract over the concrete. Black is not an easy man to know. By his own admission, "I am a man of actions, not words" (335), and his manner, seemingly aloof and demanding, often intimidates those who meet him. As he leaves Agate Hill concluding his first visit there to check on Molly, she acknowledges that she could not speak to tell him goodbye: "I said nothing, for I was terrified" (109). Although he will thereafter carry out his promise to Charlie Petree, to look after his family—now reduced to Molly—she will not make his job easy. When he comes to take her to Gatewood Academy, she clings to Selena and concedes to go only after

Selena directs her to, for though Molly recognizes his commitment to manage and pay for her education, she feels that "there is something awful about being beholden" (192). Agnes Rutherford, who has come to help bring Molly to Gatewood, states of Black, "His is an elaborately polite form of coercion and commandeering, yet his manner is such that no one could term it so" (138). She seems to appreciate his "charm" (138), as she terms his influence, as does Jacky Jarvis's cousin, B.J. Black's purchase of land up on Plain View where the Jarvises live serves as another investment in Molly's interests, as Black's visits there allow him to check both on his land and on Molly, which irritates her to the point of mean behavior. B.J. offers his more positive perspective on Black: "Mister Black? Well, he bought the mountain, of course, but he never done nothing with it. Fixed up the cabin some...but never did mix with us none...At first, Clara swore it spooked her...but it didn't spook me none...Fact is, I liked it when Mr. Black was over there, I felt like he was watching over us, or something" (292). B.J.'s portrayal reflects common notions of God the Father; Agnes and B.J. both seem to attribute almost supernatural power to Black.

Similarly, Mariah Snow, Agnes's sister and Gatewood's ascetic headmistress, sees Black as the Devil himself. Although she is maddened by his power as a major contributor to the school, she projects him as her seducer at her moment of near-breakdown under the pressures (partly self-imposed) of her position as pious woman of God. "After all these years of effort, God has sent me finally his Arch-fiend...He tempts me, Lord, yes, he tempts me" (200). When she finds that Black has no such intentions to steal her away from her dreadful post, her tone reveals disdain mixed with disappointment: "Thank God that I myself have been spared his dark attentions" (204). While her image of him is not God the Father, it nonetheless imbues him with powers spiritual and mysterious.

Molly, who is the object of Black's attentions, must only guess his motivations, in light of his solitary ways. Her lack of understanding causes her a common pitfall—she substitutes an imagined abstract meaning for experience and real knowledge. She dreads his appearances and often treats him with anger when he does show up to check on her. She does not seem to perceive him as an agent of God or of darker spiritual forces, yet he does represent to her the power of patriarchy to rob her of her independence.

The transformation of her perception, and Molly's final maturation, occurs only after the death of Jacky, to whom she gave all of her heart. As Molly grieves Jacky's death and those of the seven babies born to him and Molly, Henry arrives to tell her of Black's illness. She is surprised to hear that her benefactor is at Agate Hill, which he purchased years ago, unbeknownst to her. In a state of sudden clarity brought on by her recent losses, she decides to go to him. Her moment of leaving Plain View to return to Agate Hill is a turning point. Having come through the fire, both the literal fire that consumed Jacky's body and the metaphorical fire of their intense relationship, she is changed on some level. She considers the house she shared with Jacky and recognizes that despite the memories of their life that are so alive there, the "house …was not mine really, any more than it had ever been ours, any more than a person can lay claim to any place, for we are only passing through" (324). As the carriage pulls away from Plain View, her perspective on the scene is broader than it has ever been before; she feels that she can see "the whole wide curve of the earth" (327). From this place, she reflects, "love lives not in places nor even bodies but in the spaces between them, the long and lovely sweep of air and sky, and in the living heart and memory until that is gone too, and we are all of us wanderers, as we have always been, upon the earth." (328)

This new understanding brings to fruition the seed of contingency observable in Molly's perspective since her adolescence, and now, in its full bloom, it enables her to experience Simon Black for what he really is, a man who has loved her all these years, grateful for a purpose as her benefactor. Lying next to him as he dies, she has sloughed off all symbolic notions of him. She "lay down beside him, a thing I could never have imagined doing in all my life…I stayed with him until his death, flesh to flesh, bone to bone, pressing my body against his, the whole long fragile length of him, for it meant nothing now, and everything" (333-4). The letter Black leaves her upon his death—explaining his childhood love for Alice, his existential crisis at the destruction he observed and helped to perpetrate as a Confederate soldier, his subsequent venture in Brazil (the fortune it brought him as well as the terrible loss of his native wife and young children), his acquisition of Henry as a charge, servant, and friend, and his return to take up his responsibilities as Molly's benefactor—serves to reinforce her new recognition of him as a human being tied to material circumstances of time and space. Molly's final transformation confirms for the reader Smith's implication that meaningful growth can only be achieved by lowering the shield of abstract ideals and symbols to engage reality through experience.

With this change in her relationship to Black, effected in part by the end of his "watch" over her, Molly begins the last phase of her life, one characterized by great liberation. She tends the garden and goes to market with the now aging Henry and the blind but "gifted" Junie. Their identities are contingent on the moment: "I am the one who tells the stories, Henry is the one who drives the car, and Juney is the one who holds the basket of eggs still warm in his lap" (359). Her decentered existence embraces the beauty that can be experienced only by ridding oneself of the illusions that obscure it.

In Tuscany's closing letter to Dr. Ferrell, we are reminded that not only is Molly's reality contingent on time, place, and circumstance, but the meaning of the contents of Molly's box is contingent, as well. Tuscany interprets the items in the box to create the narrative, a narrative that she can relate to, given her own situation. Molly's life, as Tuscany experiences it, implies that Tuscany should let go of her rigid expectations if she is to open up to an authentic experience and the growth that can ensue as a result. Before his death, in the face of Aunt Cecelia's insistence that he must not let Agate Hill be "lost" to Selena, Uncle Junius says to his sister, "Frankly Cecelia Agate Hill is nothing but an encumbrance and a monument to the colossal vanity of men who enslaved other men. Let it go" (83). Through a novel rich with lessons in contingency, Smith conveys this same message to her reader.

Afterword

I would really always prefer to call what I do fiction…I think I have a better chance of being true if I call it fiction. (Lee Smith qtd. in Parrish, "Ghostland" 44)

Lee Smith's instincts for the connections among experience, story, and truth are sharp. As seen in the novels explored here, she is aware of the challenges for the contemporary subject attempting to live meaningfully, grow, and understand in a world where, on one end of the spectrum, communities can seem increasingly transient, and on the other, ideals are often substituted for a more relevant sort of truth. Smith's characters resonate, in large part, because they are pursuing an identity that not only will provide some stability but more importantly will foster growth and expansion as well as authentic connections with other people. Smith's assertion that fiction is more truthful than other kinds of stories, such as those forwarded by history or ethnographic study, emphasizes her belief that reality is slippery and yet that recognition of its protean quality is necessary to finding and understanding truth.

To accurately reflect Smith's views, we must note that in her novels, there are grave risks to ignoring or rejecting one's past: History *is* important. Katie Cocker, of *The Devil's Dream*, suffers throughout much of her journey to self-discovery due to her unwillingness to own her past, to acknowledge real, and somewhat problematic, connections with her mountain experiences as well as with her family. Her experimentation with a variety of personae, and the injuries it causes her along the way, illustrates the pitfalls of contemporary disorientation, in part because of the diminishing regional identities that once provided people with a strong sense of who they were. Yet, as I have attempted to prove here, for Smith, the more dangerous threat to the contemporary individual is the opposite force. *On Agate Hill*'s Mariah Snow is a potent example of what can happen when one's identity is defined too rigidly by traditional ideologies.

Taking seriously her obligations as a Southern Christian wife, mother, and educator of girls, Mariah is often dangerously close to breakdown: "I am locked in a golden chest, I am bound round and round by a silken rope...Nobody should trust me! For I am filled with the most base and contradictory impulses, no matter how I struggle to be worthy of God's love, and do His bidding in this world, and live up to my Responsibilities" (152-3). It is not uncommon for Smith's protagonists to express that they are "bad," recognizing their predilections to act against the expectations of family and society, expectations often conveyed through carefully shaped versions of history. However, those Smith characters who live a rich life in spite of these feelings of guilt do so by finding an identity which, unlike that posited by traditional histories, allows for contradictions and/or provides a flexibility that takes into account complex and changing circumstances. Unlike Mariah Snow, *Saving Grace*'s Gracie Shepherd finds liberation in the thought that she is beyond redemption and so is free to explore the world until she must face her inevitable consequences. She does not deny the accuracy of the history/ideology she has been taught, but she dismisses its code of behavior as one that can help her any longer. Similarly, Molly Petree, of *On Agate Hill*, tells Jacky Jarvis that she is "bad" (260), ostensibly to warn him off before he gets too seriously involved with her, but perhaps more functionally, to destroy any expectations he might have that she will be governed by common social codes. Her choice to be "bad" is a choice to subvert the restrictions of those social codes and to really *live*.

Lest readers suspect that Smith's fictional universe has abandoned itself to complete relativism, the sort that rejects any system of ethics and/or healthy community, never fear. Like her implication that we must not dismiss history altogether but instead should adopt a sense of past that allows for contingency, Smith posits that ethics and community are central to healthy and meaningful existence. As expressed in Chapter 9 of this study, Smith does not promote a kind of relative truth that justifies *any* and *all* actions. For example, *On Agate Hill*'s Nicky Eck is portrayed as despicable for his molestation of the young Molly Petree when she is too young to know how to defend herself against his advances. He is dealt with by the community itself, first stabbed in the back with a pitchfork by Spencer Hall, out of a desire to protect Molly, and later killed by someone unknown to Molly, ostensibly for similar infractions--crimes defined as such, again, by the community affected. This reaction by the community is evidence of John Rothfork's assertion that postmodern relativism does allow for ethics, and that an attitude of contingency actually provides a more effective ethics than idealism does. Consistent with this attitude,

Smith critiques the opposite extreme, codes of behavior that not only dictate how one must behave in all situations, but posit absolute moral meanings by which these human behaviors must always be judged.

In carving out space for identities relevant to the contemporary subject, Smith claims certain narrative tools to great effect. As we see in the works represented here, in order to convey the necessity of fluidity and contingency for a rich and authentic experience, she utilizes (fictional) oral histories translated into written story, letters, diary entries, multiple-voice narratives, songs, folk stories and poems, as well as ostensible anthropological studies. Those who fare worst in Smith's works are characters who 1) cannot break free psychologically from traditional definitions of self and morality, like Mariah Snow, and/or 2) retreat from the messiness of real life to a safe and buffered existence, like Lizzie Bailey in *The Devil's Dream*, who chooses the life of a war nurse, antiseptic and solitary, over the "fire" of intense engagement with other people. Through varied approaches to narrative, Smith not only explores the characters' identity-pursuits, but often also illustrates that complex process through the narrative itself. For example, we are confronted throughout the metafictional *On Agate Hill* with the fact that in processing reality, we must constantly interpret, as Tuscany Miller does. Smith emphasizes, through her use of forms, that this ever-interpretive process is inextricably linked to our experience of *self*.

As this book goes to press, Smith is undoubtedly writing a new novel, so further study of her work is definitely warranted. Not only does her short fiction merit continued critical attention, but as the body of her work grows, comprehensive study of that entire oeuvre will help illuminate Smith's exploration of identity. The full impact of her commentary on this subject is still to be seen, but it is clear even now that there is much to be gained by her fictional insights into contemporary experience, particularly in the context of the American South.

Bibliography

Primary Sources:

Smith, Lee. *Black Mountain Breakdown*. New York, NY: Ballantine, 1980.

---. *Cakewalk*. New York, NY: Putnam, 1981.

---. *The Christmas Letters*. Chapel Hill, NC: Algonquin, 1996.

---. *The Devil's Dream*. New York, NY: Putnam, 1992.

---. *Fair and Tender Ladies*. New York, NY: Ballantine, 1988.

---. *Fancy Strut*. New York, NY: Harper and Row, 1973.

---. *Family Linen*. New York, NY: Ballantine, 1985.

---. *The Last Day the Dogbushes Bloomed*. New York, NY: Harper and Row, 1968.

---. *The Last Girls*. Chapel Hill, NC: Algonquin, 2002.

---. *Me and My Baby View the Eclipse*. New York, NY: Ballantine, 1990.

---. *Mrs. Darcy and the Blue-Eyed Stranger*. Chapel Hill, NC: Algonquin, 2011.

---. *News of the Spirit*. New York, NY: Ballantine, 1997.

---. *On Agate Hill*. Chapel Hill, NC: Algonquin, 2006.

---. *Oral History*. New York, NY: Ballantine, 1983.

---. *Something in the Wind*. New York, NY: Harper and Row, 1971.

---. *Saving Grace*. New York, NY: Ballantine, 1995.

Secondary Sources:

Barthes, Roland. *The Pleasure of the Text*. Trans. Richard Miller. New York, NY: Hill and Wang, 1973.

Bower, Anne Lieberman. "Rewriting the Self, Writing the Other: An Investigation of Recent American Epistolary Novels." Diss. West Virginia U, 1990.

Buchanan, Harriette C. "Lee Smith: The Storyteller's Voice." *Southern Women Writers: The New Generation*. Ed. Tonette Bond Inge. Tuscaloosa: U of Alabama P, 1990. 324-371.

Byrd, Linda. "The Emergence of the Sacred Sexual Mother in Lee Smith's *Oral History*." *Southern Literary Journal* 31.1 (Fall 1998): 119-143.

Byrd-Cook, Linda J. "Reconciliation with the Great Mother Goddess in Lee Smith's *Saving Grace*." *Southern Quarterly* 40.4 (Summer 2002): 97-112.

Campbell, H.H. "Lee Smith and the Bronte Sisters." *Southern Literary Journal* 33.1 (Fall 2000): 141-49.

Cunningham, Rodger. "Writing on the Cusp: Double Alterity and Minority Discourse in Appalachia." *The Future of Southern Letters*. Ed. Jefferson Humphreys and John Lowe. New York, NY: Oxford UP, 1996. 41-53.

Derrida, Jacques. *Of Grammatology*. Trans. Gayatri Chakravorty Spivack. Baltimore, MD: Johns Hopkins, 1974.

Donlon, Joycelyn Hazelwood. "Hearing Is Believing: Southern Racial Communities and Strategies of Story-Listening in Gloria Naylor and Lee Smith." *Twentieth Century Literature* 41.1 (Spring 1995): 16-35.

Donoghue, Denis. "Deconstructing Deconstruction." *Critical Essays on American Postmodernism*. Ed Stanley Trachtenberg. New York, NY: G.K. Hall, 1995. 31-4.

Doyle, Jacqueline. "'These Dark Woods Yet Again': Rewriting Redemption in Lee Smith's *Saving Grace*." *Critique* 41.2 (Spring 2000): 273-290.

Eckard, Paula Gallant. *Maternal Body and Voice in Toni Morrison, Bobbie Ann Mason, and Lee Smith*. Columbia: U of Missouri P, 2002.

---. "The Prismatic Past in *Oral History* and *Mama Day*." *MELUS* 20.3 (Fall 1995): 121-35.

Faulkner, William. *Light in August*. New York, NY: Vintage, 1985.

Flax, Jane. *Thinking Fragments: Psychoanalysis, Feminism, and Postmodernism in the Contemporary West*. Berkeley: U of California P, 1990.

Gordon, David J. "Iris Murdoch's Comedies of Unselfing." *Twentieth Century Literature* 36.2 (1990): 115-136.

Gwin, Minrose. "Nonfelicitous Space and Survivor Discourse." *Haunted Bodies: Gender and Southern Texts*. Eds. Anne Goodwyn Jones and Susan V. Donaldson. Charlottesville: U of Virginia P, 1997. 416-440.

Hall, Joan Wylie. "Arriving Where She Started: Redemption at Scrabble Creek in Lee Smith's *Saving Grace*." *Pembroke Magazine* 33 (2001): 80-85.

Hassan, Ihab. "Pluralism in Postmodern Perspective." *Critical Inquiry* 12.3 (Spring 1986): 503-520.

Hentchel, Uwe, Juris D. Draguns, Wolfrum Elhers, and Gudmund Smith. "Defense Mechanisms: Current Approaches to Research and Measurement." *Defense Mechanisms (Volume 136): Theoretical Research, and Clinical Perspectives*. Eds. Uwe Hentchel, Gudmund Smith, Juris D. Draguns, and Wolfrum Elhers. Amsterdam: Elsevier B.V., 2004. 3-42.

Herion-Sarafidis, Elisabeth. Interview with Lee Smith. *Southern Quarterly* 32.2 (1994): 7-18.

Hill, Dorothy Combs. *Lee Smith*. New York, NY: Twayne, 1992.

Hobson, Fred. *The Southern Writer in the Postmodern World*. Athens: U of Georgia P, 1991.

Jameson, Fredric. *Postmodernism, Or, The Cultural Logic of Late Capitalism*. Durham, NC: Duke UP, 1991.

Jones, Anne Goodwyn. "The World of Lee Smith." *Women Writers of the Contemporary South*. Ed. Peggy Whitman Prenshaw. Jackson: UP of Mississippi, 1984. 249-72.

Jones, Suzanne W. "City Folks in Hoot Owl Holler: Narrative Strategy in Lee Smith's *Oral History*." *Southern Literary Journal* 20.1 (1987): 101-112.

Kalb, John D. "The Second 'Rape' of Crystal Spangler." *Southern Literary Journal* 21.1 (1988): 23-30.

Kearns, Katherine. "From Shadow to Substance: The Empowerment of the Artist Figure in Lee Smith's Fiction." *Writing the Woman Artist: Essays on Poetics, Politics, and Portraiture*. Ed. Suzanne Whitmore Jones. Philadelphia: U of Penn. P, 1991. 175-95.

Kingston, Maxine Hong. *The Woman Warrior: Memoirs of a Girlhood Among Ghosts*. New York, NY: Vintage Books, 1976.

Kline, Paul. "A Critical Perspective on Defense Mechanisms." *Defense Mechanisms (Volume 136): Theoretical Research, and Clinical Perspectives*. Eds. Uwe Hentchel, Gudmund Smith, Juris D. Draguns, and Wolfrum Elhers. Amsterdam: Elsevier B.V., 2004. 43-54.

Kreyling, Michael. *Inventing Southern Literature*. Oxford, MS: UP of Mississippi, 1998.

Ladd, Barbara. "Literary Studies: The Southern United States, 2005." *PMLA* 120.4 (Oct. 2005): 1628-1639.

MacKethan, Lucinda. "Artists and Beauticians: Balance in Lee Smith's Fiction." *Southern Literary Journal* 15.1 (1982): 3-14.

---. *Daughters of Time: Creating Woman's Voice in Southern Story*. Athens: U of Georgia P, 1990.

Massey, Kevin. "'Wonderful Terms and Phrases': Contrasting Dialect in William Faulkner's *As I Lay Dying* and Lee Smith's *Oral History*." *North Carolina Literary Review* 7 (1998): 11-19.

Merton, Thomas. *Zen and the Birds of Appetite*. New York, NY: New Directions, 1968.

Moskowitz, Andrew, and Ceri Evans. "Peritraumatic Dissociation and Amnesia in Violent Offenders." *Dissociation and the Dissociative Disorders: DSM-V and Beyond*. Eds. Paul F. Dell and John A. O'Neil. New York, NY: Routledge, 2009. 197-208.

Parrish, Nancy C. "'Ghostland': Tourism in Lee Smith's *Oral History*." *Southern Quarterly* 32.2 (1994): 37-47.

---. *Lee Smith, Annie Dillard, and the Hollins Group: A Genesis of Writers*. Baton Rouge: Louisiana State UP, 1998.

Reilly, Rosalind B. "*Oral History*: The Enchanted Circle of Narrative and Dream." *Southern Literary Journal* 23.1 (Fall 1990): 79-92.

Rothfork, John. "Postmodern Ethics: Richard Rorty and Michael Polanyi." John Rothfork (web page). North Arizona University. http://oak.ucc.nau.edu/jgr6/pubs.html. Accessed May 23, 2013. Originally published in *Southern Humanities Review* 29.1 (1995): 15-48.

Smith, Rebecca. "A Conversation with Lee Smith." *Southern Quarterly* 32.2 (1994): 19-29.

---. "Writing, Singing and Hearing a New Voice: Lee Smith's *The Devil's Dream*." *The Southern Quarterly* 32.2 (1994): 48-62.

Steinberg, Marlene. "Systematizing Dissociation: Symptomatology and Diagnostic Assessment." *Dissociation: Culture, Mind and Body*. Ed. David Spiegel. Washington D.C.: American Psychiatric P, 1994. 59-88.

Suzuki, D.T. *Zen Buddhism*. Ed. William Barrett. Garden City, N.Y.: Doubleday, 1956.

Town, Caren J. *The New Southern Girl: Female Adolescence in the Works of 12 Women Authors*. Jefferson, NC: McFarland, 2004.

Trachtenberg, Stanley. Introduction. *Critical Essays on American Postmodernism*. Ed. Stanley Trachtenberg. New York, NY: G.K. Hall, 1995. 1-27.

Tucker, Lindsey. Introduction. *Critical Essays on Iris Murdoch*. Lindsey Tucker, ed. New York, NY: G.K. Hall, 1992. 1-16.

Watts, Alan W. *The Way of Zen*. New York, NY: Vintage Books, 1957.

Wesley, Debbie. "A New Way of Looking at an Old Story: Lee Smith's Portrait of Female Creativity." *Southern Literary Journal* 30.1 (1997): 89-101.

Winchell, Mark Roydon. *Reinventing the South: Versions of a Literary Region*. Columbia, MO: U of Missouri P, 2006.

Yaeger, Patricia. *Dirt and Desire: Reconstructing Southern Women's Writing*, 1930-1990. Chicago, IL: U of Chicago P, 2000.

Tanya Long Bennett is a professor of English at University of North Georgia, where she has taught for thirteen years. She earned her PhD in English at University of Tennessee. Her research focuses on twentieth and twenty-first century fiction, as well as gender studies.

Index

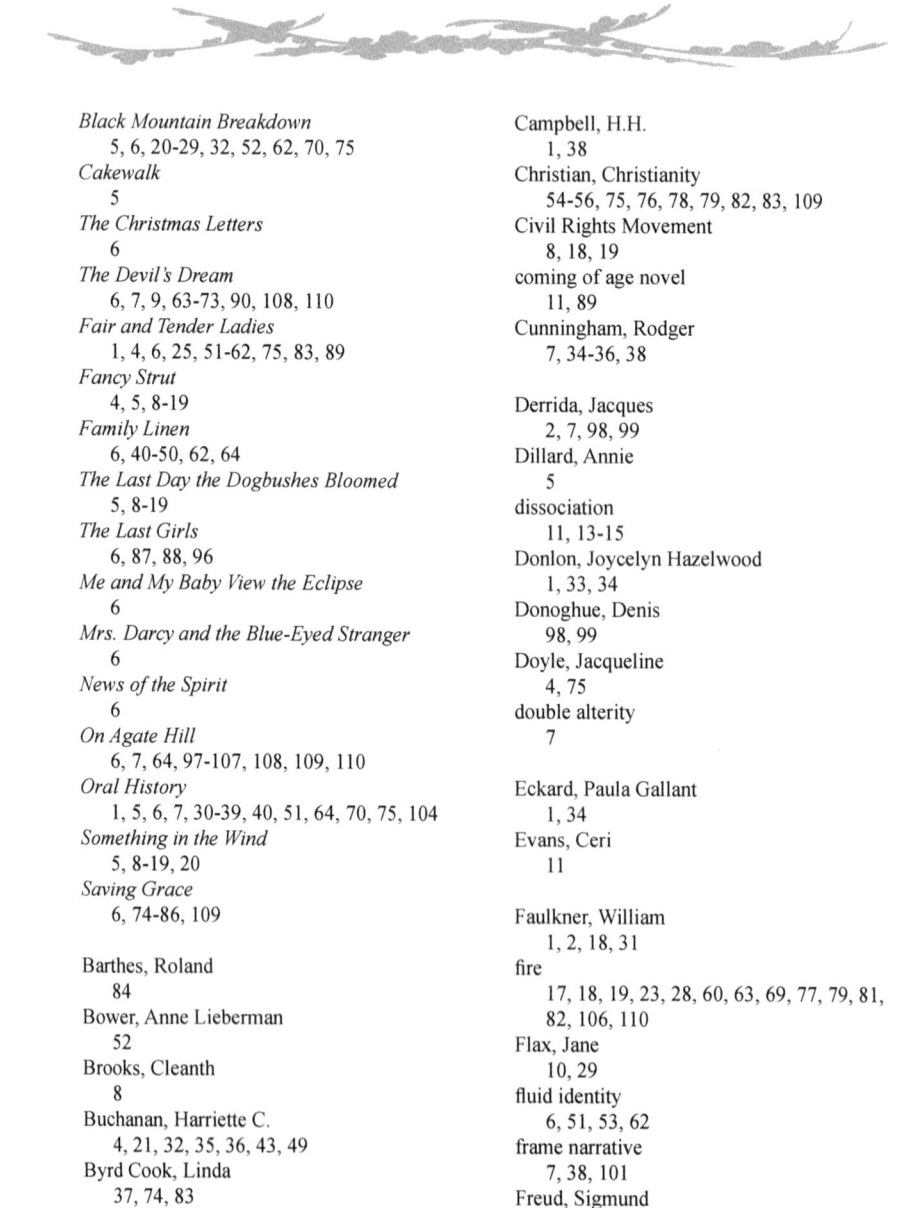

Black Mountain Breakdown
   5, 6, 20-29, 32, 52, 62, 70, 75
Cakewalk
   5
The Christmas Letters
   6
The Devil's Dream
   6, 7, 9, 63-73, 90, 108, 110
Fair and Tender Ladies
   1, 4, 6, 25, 51-62, 75, 83, 89
Fancy Strut
   4, 5, 8-19
Family Linen
   6, 40-50, 62, 64
The Last Day the Dogbushes Bloomed
   5, 8-19
The Last Girls
   6, 87, 88, 96
Me and My Baby View the Eclipse
   6
Mrs. Darcy and the Blue-Eyed Stranger
   6
News of the Spirit
   6
On Agate Hill
   6, 7, 64, 97-107, 108, 109, 110
Oral History
   1, 5, 6, 7, 30-39, 40, 51, 64, 70, 75, 104
Something in the Wind
   5, 8-19, 20
Saving Grace
   6, 74-86, 109

Barthes, Roland
   84
Bower, Anne Lieberman
   52
Brooks, Cleanth
   8
Buchanan, Harriette C.
   4, 21, 32, 35, 36, 43, 49
Byrd Cook, Linda
   37, 74, 83

Campbell, H.H.
   1, 38
Christian, Christianity
   54-56, 75, 76, 78, 79, 82, 83, 109
Civil Rights Movement
   8, 18, 19
coming of age novel
   11, 89
Cunningham, Rodger
   7, 34-36, 38

Derrida, Jacques
   2, 7, 98, 99
Dillard, Annie
   5
dissociation
   11, 13-15
Donlon, Joycelyn Hazelwood
   1, 33, 34
Donoghue, Denis
   98, 99
Doyle, Jacqueline
   4, 75
double alterity
   7

Eckard, Paula Gallant
   1, 34
Evans, Ceri
   11

Faulkner, William
   1, 2, 18, 31
fire
   17, 18, 19, 23, 28, 60, 63, 69, 77, 79, 81, 82, 106, 110
Flax, Jane
   10, 29
fluid identity
   6, 51, 53, 62
frame narrative
   7, 38, 101
Freud, Sigmund
   10

Gordon, David J.
  100
Gwin, Minrose
  21

Hall, Joan Wylie
  75
Hassan, Ihab
  9
Hentchel, Uwe
  10
Herion-Sarafidis, Elisabeth
  52, 53
Hill, Dorothy Combs
  4-6, 52, 53 ,56
Hobson, Fred
  2, 31
historical ideology
  2
history
  1, 2, 4, 6, 8, 9, 18, 27, 30, 31, 33, 35, 38-40, 44, 47, 61, 63-65, 69, 71, 72, 75, 88, 96, 99, 100, 103, 104, 108, 109
  family history
    17, 33, 40-46, 94
  folk history
    38, 38
  historical identity
    30, 32, 34, 42, 71
  historical meditation
    3, 31
  oral history
    5, 33-35, 110
Huck Finn
  58, 87, 88, 89, 93

Ivy Rowe
  1, 4, 25, 34, 51-62, 83, 89

Jameson, Fredric
  4
Jones, Anne Goodwyn
  5, 6, 22, 23
Jones, Suzanne W.
  4, 35

Kalb, John D.
  24
Kearns, Katherine
  5, 14, 16, 49, 61
Kingston, Maxine Hong
  73
Kline, Paul
  11
Kreyling, Michael
  8, 9

Ladd, Barbara
  3, 4

MacKethan, Lucinda
  5, 16, 23, 24, 29, 60, 61
Massey, Kevin
  1
Merton, Thomas
  7, 64, 65, 73
mirror
  5, 25, 26, 38, 51, 53, 60, 61

Morrison, Toni
  1
Moskowitz, Andrew
  11
multiple points of view
  16, 87, 88

Naylor, Gloria
  1

O'Connor, Flannery
  31, 75

Parrish, Nancy C.
  5, 61, 69, 108
postmodern subject
  2, 4, 7, 64
postmodernism
  2, 3, 7-10, 19, 21, 28, 31, 38, 51, 62, 64, 79, 99, 100, 109
primary experience
  63-65, 68-70, 72, 73, 79
Protean
  29, 51, 55, 59, 60, 62, 83, 84, 86, 108

Reilly, Rosalind B.
  38
Rothfork, John
  100, 101, 109
Rubin, Louis D.
  5, 8

secondary experience
  7, 63, 65-69
Smith, Rebecca
  2, 66, 71
split self
  10, 11, 14, 19
Southern Renascence
  3, 8
Steinberg, Marlene
  11
Suzuki, D.T.
  64, 65

Tate, Allen
  3
Town, Caren J.
  14
Trachtenberg, Stanley
  7, 98, 99
Tucker, Lindsey
  100

Watts, Alan W.
  7, 70, 71
Wesley, Debbie
  49, 72
Winchell, Mark Royden
  3

Yaeger, Patricia
  4, 8, 9, 21, 104

Zen (and Zen Buddhism)
  7, 64, 65, 70, 71, 73